Recent Advances in Pollution Control by Treatment Wetlands

In Focus – Special Book Series

Recent Advances in Pollution Control by Treatment Wetlands

Editors

Bernhard Pucher and Guenter Langergraber

Published by

**IWA Publishing
Unit 104-105, Export Building
1 Clove Crescent
London E14 2BA, UK**
Telephone: +44 (0)20 7654 5500
Fax: +44 (0)20 7654 5555
Email: publications@iwap.co.uk
Web: www.iwaponline.com

First published 2024
© 2024 IWA Publishing

Disclaimer

British Library Cataloguing in Publication Data
A CIP catalogue record for this book is available from the British Library

ISBN: 9781789064872

Contents

doi: 10.2166/wst.2023.196

Editorial: Recent advances in pollution control by treatment wetlands (WETPOL 2021)

This book includes selected contributions presented at the '9th International Symposium on Wetland Pollutant Dynamics and Control (WETPOL 2021)' that was organized by the University of Natural Resources and Life Sciences, Vienna (BOKU) and held as a virtual event from 13 to 17 September 2021. WETPOL is an international symposium bringing together wetland scientists, engineers, and practitioners working on wetland ecosystem services, including water quality improvement, climate regulation, and flood control. The goal of WETPOL 2021 was to improve the understanding of the role wetlands perform in processing nutrients and contaminants, and to discuss and demonstrate how restored and constructed wetlands in the future, via their associated ecosystem services, can contribute to ensure sustainable water management and resource recovery while at the same time regulating and mitigating impacts of global climate change.

In total, more than 200 participants from 38 countries participated virtually in WETPOL 2021. One hundred and thirty-three talks and 29 posters were presented within the 16 general sessions and 11 submitted special sessions. The topics with the most contributions included nature-based solutions in the urban environment as well as wetlands for agricultural drainage and diffuse pollution control. In a virtual field trip, the participants were able to experience nature-based solution from Vienna, implemented treatment wetlands in Austria and the research facility Wassercluster Lunz.

This book includes nine contributions that were presented at WETPOL 2021. These contributions show the variety of topics presented at the conference. The topics include experimental work and simulation studies regarding

- various treatment wetland types, e.g. free water surface wetlands, subsurface flow wetlands, and intensified wetlands,
- the performance of treatment wetlands for different types of water, e.g. greywater, industrial wastewater, and agricultural runoff,
- the performance of treatment wetlands for removing selected micro-pollutants, as well as
- the impact of salinity on the treatment performance.

We do hope that you enjoy reading the selected contributions from WETPOL 2021.

Guest Editors
Bernhard Pucher **IWA**
Guenter Langergraber **IWA**
University of Natural Resources and Life Sciences, Vienna, Department of Water, Atmosphere and Environment, Institute of Sanitary Engineering and Water Pollution Control, Muthgasse 18, 1190 Vienna, Austria

doi: 10.2166/wst.2022.184

Reciprocating subsurface-flow constructed wetlands: a microcosm treatability study

Leslie Behrends [a,b,†,*], Laura Houke[a,†], Earl Bailey[a,†] and Pat Jansen[a,†‡]

[a] Department of Air, Land and Water Sciences, Tennessee Valley Authority (TVA), Muscle Shoals, AL, USA
[b] Reciprocating Wetlands Water Technologies, 1070 Goshentown Road, Hendersonville, TN 37075, USA
*Corresponding author. E-mail: leslielbehrends@yahoo.com

LB, 0000-0002-7514-0329

ABSTRACT

Two batch-loaded microcosm treatability studies of eight days' duration (192 hours) were conducted concurrently from July 26 through August 3 in an environmentally controlled greenhouse at the Tennessee Valley Authority's (TVA) Constructed Wetlands Research Facility in Muscle Shoals, Alabama, USA. These, the first of five treatability studies, were conducted in batch-loaded wetland microcosms, with and without reciprocation. Reciprocation refers to the process of filling and draining paired wetland cells on a recurrent and timed basis to facilitate passive aeration of fixed microbial biofilms during the drain phase and oxygen depletion during the fill phase. The study was designed to simultaneously evaluate the influence of three treatment parameters with respect to removal dynamics of chemical oxygen demand (COD), total ammonia-nitrogen (TAN), nitrate-nitrogen, and orthophosphorus. The treatment parameters evaluated included four COD concentrations, two reciprocation cycle times, and the presence or absence of reed canary grass (*Phalaris arundinacea*), a plant with an anoxic rootzone. Temperature, pH, dissolved oxygen, electrical conductivity, redox potential, and evapotranspiration were also monitored. Results showed that treatments with reciprocation vs. no reciprocation provided significantly enhanced aerobic and anoxic treatment of wastewater and improved removal of COD and ammonium, irrespective of initial COD concentrations or planted vs. not planted with *P. arundinacea*. Treatments with *P. arundinacea* had less accumulation of nitrate nitrogen; less than 0.4 mg/L., while reciprocating treatments without *P. arundinacea* accumulated from 15–26 mg/L nitrate nitrogen over the 192-hour treatment period.

Key words: denitrification, nitrification, reciprocating wetlands, redox-potential, subsurface-flow constructed wetlands

HIGHLIGHTS

- Reciprocation improved several aerobic and anoxic wastewater treatment processes.
- Reciprocation improved nitrification.
- Reciprocation enhanced denitrification.
- Canary grass, *Phalaris arundinacea*, improved denitrification.
- Reciprocation has potential to be energy efficient and land conserving as compared to traditional subsurface flow constructed wetlands.

† Retired from TVA.

‡ deceased.

GRAPHICAL ABSTRACT

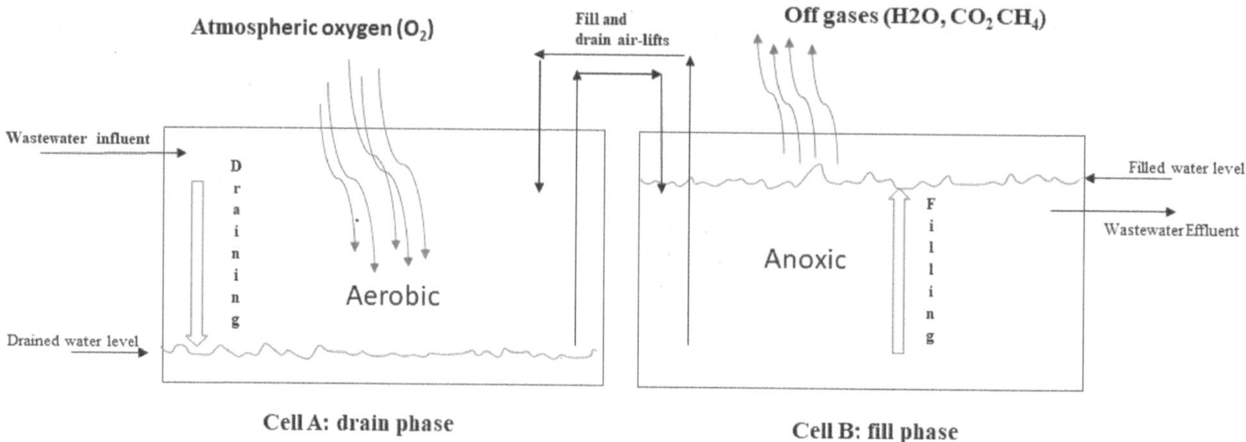

INTRODUCTION

Chronically low dissolved oxygen (D) concentrations in subsurface-flow constructed wetlands (SSFW) adversely impact aerobic treatment processes including organic matter degradation and nitrification (Vymazal & Kropfelova 2008; Garcia *et al.* 2010). Oxygen dynamics in gravel-based SSFW are affected by saturated media profiles, low D solubility (8–16 mg/L), low convective mixing, negligible diffusion of oxygen at the air-water interface, and high community respiration rates. Net transfer of D via diffusion at the air-water interface in SSFW is marginal, ranging from 0.2 to 3.2 g/m^2/day (Behrends *et al.* 1996; Tyroller *et al.* 2010; Nivala *et al.* 2013).

To overcome this problem, various ideas and concepts have been evaluated to enhance oxygen transfer in SSFW. García *et al.* (2005) obtained significant improvements in wastewater treatment by reducing the depth of the wetland substrate from 50 to 25 cm. However, this method requires doubling the land area requirements for a given hydraulic retention time, which adds to land and capital costs. Data suggests that aquatic macrophytes may or may not contribute significant D to the rhizosphere in SSFW, with results ranging from 0.005 to 12 g/m^2/day (Brix & Schierup 1990; Steinberg & Coonrod 1994; Armstrong *et al.* 2000). However, based on empirical evidence gathered from field trials, other researchers have concluded that the amount of oxygen made available to the rhizosphere by aquatic macrophytes is at least an order of magnitude less than the amount needed for aerobic treatment of domestic and municipal wastewater (Nivala *et al.* 2013).

While oxygen diffusion at the air-water interface is rate limited as discussed above, oxygen diffusion in the atmosphere is up to 10,000 times faster. So Brix & Schierup (1990) and Watson & Danzig (1993) recommended intermittent flooding and draining of SSFW cells to improve atmospheric oxygen supply to plant roots and substrate biofilms. Green *et al.* (1997), discussed and proved the utility of a passive air-pump to enhance nitrification in gravel-based wetlands. Others have used the fill and drain process to improve oxygenation of plant roots and to significantly improve nitrification in aquaponic food production systems (Rakocy 2012).

Starting in 1993, TVA scientists began research and development of a novel reciprocating wetland, whereby coupled SSFW treatment cells were recurrently filled and drained (six to 12 times/day), such that dewatered substrate biofilms and plant roots were repeatedly exposed to atmospheric oxygen (Behrends 1999). During the dewatered drain phase, the roots and biofilms were exposed to a 21% oxygen atmosphere (300 mg/L O_2), for up to 2 hours per cycle. During the fill phase, these same roots and biofilms were exposed to an oxygen-deprived environment (anoxic), which promoted enhanced removal of nitrate via microbial denitrification. This fill and drain reciprocating process enhanced growth of select aquatic macrophytes, COD removal, and nitrification and denitrification (Behrends *et al.* 1996). Variations and modifications of the TVA reciprocating system have been referred to as fill and drain wetlands and tidal-flow wetlands (Austin & Nivala 2009; Wu *et al.* 2011).

To reiterate, the role of reciprocation is *not* to aerate the wastewater per se, but instead, it is a method for supplying an ever abundant and low-cost supply of atmospheric oxygen to the microbial and plant biota during the reciprocation drain phase. As will be verified in this batch-loaded microcosm study, oxygen dynamics, pH, and redox potential of paired SSFW systems can be controlled and optimized via reciprocation to enhance both aerobic and anoxic wastewater treatment processes. Discussions will be developed around these microcosm-based studies as well as later findings from pilot and demonstration scale reciprocating systems.

METHODS

Two microcosm-based treatability studies were conducted concurrently to evaluate the reciprocation process and to quantify treatment dynamics and interactions over a 196-hour trial. Variables monitored included DO, temperature (°C), pH, redox potential (mV), electrical conductivity (EC), chemical oxygen demand (COD), total ammonia nitrogen (TAN), nitrate nitrogen (NO3-N), orthophosphorus (PO$_4$-P), and evapotranspiration (ET).

Powdered milk replacement starter (MRS) was used in these studies as an organic carbon source for producing synthetic wastewater. It is highly soluble and can be used to craft various strengths of synthetic wastewaters for treatability studies. MRS has a complex mix of protein, fats, minerals, and micronutrients that simulate wastewater and help to sustain microbial biofilms and aquatic plants (Table 1). Freshly prepared MRS solutions showed a high correlation for well-mixed solutions having 50, 100, 200, and 400 mg/L MRS dry powder:

$$(y = 0.4144x + 3.9258, R^2 = 1, \text{ where } y = BOD_5 \text{ and } x = COD).$$

Table 1 summarizes information related to the composition of MRS, while Table 2 provides treatment designations, nutrient treatments, reciprocation regimes, and planting arrangements (plants vs. no plants).

Microcosm units

Microcosm units consisted of 30 opaque glass aquaria (70 litres volume, 60 cm × 38 cm × 30 cm). Each microcosm had a surface area equivalent to 0.228 m^2 and was divided into two equal volumes with a permanently fixed water tight vertical glass partition. Each of the two compartments within a microcosm were plumbed to accommodate recurrent draining and filling (reciprocation) via independent air-lift devices (Figures 1 and 2). Both sides of each microcosm were backfilled with 2.5 cm of limestone rock (2.5–4.0 cm), which was overlaid with 25 cm of river pea-gravel (0.5 to 1.0 cm). Porosity of the rock substate averaged 40 percent. Prior to adding the gravel substrates, 38 cm lengths of slotted PVC well-screen ports (5 cm diameter) were placed vertically on either side of the center partition (Figure 2). These ports enabled taking water samples, checking water quality with multi-parameter sondes, and checking redox potential with in-situ platinum-tipped probes.

Table 1 | Composition of MRS used as a source of organic carbon and elements for preparing synthetic wastewater for batch loaded microcosm-based treatability studies

Element	Concentration (%)	Element	Concentration (mg/kg)
Organic C	43.7	Fe	179
N	3.56	Zn	47.3
P	0.74	Mn	46.7
K	2.38	Cu	13.3
Mg	0.17	Ni	2.3
Ca	0.80	Co	<0.3
Na	0.17	SO^4-S	160
Si	0.94		
S	0.32		
CO^3-C	1.52		

Table 2 | Treatment designations and treatment parameters

Treatment Designation[a]	MRS Loading (mg/L)	COD (mg/L)	TAN (mg/L)	PO4-P (mg/L)	Reciprocation Regime[b]	Planting Regime
50–0-Y	50	111	100	50	0	Y
50-1-Y	50	111	100	50	1	Y
50-2-Y	50	111	100	50	2	Y
100-0-Y	100	238	100	50	0	Y
100-1-Y	100	238	100	50	1	Y
100-2-Y	100	238	100	50	2	Y
200-0-Y	200	516	100	50	0	Y
200-1-Y	200	516	100	50	1	Y
200-2-Y	200	516	100	50	2	Y
400-0-N	400	988	100	50	0	N
400-0-Y	400	988	100	50	1	Y
400-1-N	400	988	100	50	2	N
400-1-Y	400	988	100	50	0	Y
400-2-N	400	988	100	50	1	N
400-2-Y	400	988	100	50	2	Y

Each treatment combination was replicated twice. The two treatability studies were run concurrently during the summer (July 26 to August 3).
[a]Treatment designation 50-1-Y, for example, refers to the MRS concentration (50 mg/L); reciprocation rate (1 hour); and planted with canary grass (Y) vs. not planted (N). The third column references the COD concentration (mg/L) for the respective MRS concentration.
[b]Reciprocation regimes were 0 (no reciprocation), 1 (12 cycles/day), and 2 (six cycles/day) respectively. Digital timers were programmed to start and stop air lifts based on either one- or two-hour reciprocation regimes.

Figure 1 | Glass-partitioned aquaria (70 L) were used as microcosms for batch-loaded and flow-through treatability studies. Vertical pipe-- within-pipe networks with air tubing served as air-lifts for moving water from Cell A to Cell B and vice-versa.

Figure 2 | Experimental microcosm facility illustrating planted (canary grass) and unplanted treatments. Air lift assemblies to the rear of each aquarium were powered by regenerative blowers (lower left). Frequency and duration of air-lift reciprocation cycles were controlled by programmable timers, solenoid valves, and one-way aquaria valves.

Acclimation, microbial seeding, and planting of microcosms

Acclimation of planted microcosms consisted of batch loading 13 L of a synthetic wastewater solution into one side of each microcosm. The solution had 200 mg/L MRS as a source of organic carbon, and macro- and micro- nutrients (Table 1). The synthetic wastewater was amended with 50 ml of a 'dirty' water slurry collected from an existing SSFW. The slurry was assumed to have a diverse population of native microbial population that would supply seeding of the wetland microcosms. During acclimation (two weeks prior to microcosm trials), all microcosms remained static, with no reciprocation. The two-week acclimation period was intended for biotic establishment prior to the treatability studies. Designated treatments were planted with canary grass (*Phalaris arundinacea*) at 50 g wet weight, equivalent to 219 g/m^2. Other treatments were not planted and served as non-planted controls (Table 2).

Experimental design(s) and treatment designations

Experimental designs consisted of two factorial experiments that were run concurrently. The first batch-loaded study evaluated three MRS concentrations (50,100, and 200 mg/L), equivalent to COD concentrations of 111, 238, 516 mg/L, two reciprocation treatments (six and 12 cycles/day), and a non-reciprocating control. All treatments in study one were planted with canary grass (*P. arundinacea*), a plant that has an anoxic rootzone which can enhance denitrification (Steinberg & Coonrod (1994); Zhu & Sikora (1995). The second concurrent study evaluated batch loading of a single MRS concentration (400 mg/L), equivalent to 988 mg/L COD, two reciprocation treatments (six and 12 cycles/day), a non-reciprocating control, and two planting regimes (with and without canary grass, *P. arundinacea*). In both studies, treatment combinations were replicated twice. See Table 2 footnotes regarding treatment definitions and designations for both treatability studies.

Synthetic wastewater preparation, batch-loading, and initiation of experiments

Four 200 L batches of synthetic wastewater were prepared using dechlorinated tap water, powdered MRS, ammonium chloride and potassium phosphate (Table 2). Reciprocation modes included a no-reciprocation control, a one-hour and a two-hour regime.

In the one-hour reciprocation regime water was air-lifted from Cell A to Cell B for twenty minutes followed by a 40-minute rest period. Twenty minutes of air lift was sufficient to move a 25 cm column of water in Cell A to Cell B and vice versa.

Subsequently, water was air-lifted from Cell B to Cell A for 20 minutes followed by a 40-minute rest period. The two-hour reciprocation regime consisted of air-lifting wastewater from Cell A to Cell B for 20 minutes followed by a 100-minute rest period. Subsequently, water was air-lifted from Cell B to Cell A for 20 minutes followed by a 100-minute rest period. Thus, during a 24-hour period, there were 12 and 6 cycling events for the one- and two-hour reciprocation regimes, respectively.

Wastewater management, sampling, and analyses

After the two-week acclimation period, the microcosm units were drained and refilled with 13 L of simulated wastewater according to treatment designation (Table 2). Wastewater samples were collected at this time from each of the four batches of synthetic wastewater and analyzed to determine composition of synthetic wastewater solutions at time $= 0$. Subsequently, water samples were collected from microcosm units with syringes from the vertical PVC wells at elapsed times of 1, 3, 6, 12, 24, 48, 96, and 192 hours. Duplicate samples were preserved (or not) with 2.0 mL of concentrated sulfuric acid and stored at 4 C until analyzed (usually within 48 hours). Unpreserved samples were used for COD analyses using Hach kits (Hach Chemical Co, Ames, Iowa). TAN, nitrate, and ortho-P were analyzed using either methods of inductively coupled plasma (ICP), or an automated Latchet instrument.

Water quality parameters including temperature (°C), DO, EC, and pH were also checked at 1, 3, 6, 12, 244, 896 and 192 hours with a multi-parameter hand-held sonde (YSI 600 YSI Inc., Yellow Springs, OH). Redox potential was checked with a platinum tipped copper wire that were mounted inside the PVC monitoring wells, with the platinum tip 7.5 cm from the bottom. A millivolt (mV) meter was connected to the redox probe with a calomel reference electrode. Redox values were corrected by adding 244 mV to adjust values to a hydrogen reference (Zhu & Sikora 1995).

ET was quantified by recording the amount of dechlorinated tap water that was added back to each microcosm during the study to replace water lost to either evaporation (no-plant controls), or ET in planted treatments. These data were summed by treatment to illustrate water use as a function of MRS concentration, reciprocation treatment (0 vs. six vs. 12 cycles), and planting regime (plants vs. no plants).

Data summarization and statistical analysis

Reaction rate coefficients (k) were calculated for removal of COD and TAN using the linearized first-order kinetic model: ln $(C_0/C) = kt$, where C and C_0 represent concentrations at time t and the starting concentration respectively (Crites & Tchobanoglous 1998). Means and standard deviations were calculated for tabular and graphic illustrations.

RESULTS AND DISCUSSION

Temperature, pH, and conductivity

Average temperatures (°C) among treatments over the eight-day trial ranged from 28.2 to 31.3. (Figure 3). While there were marginal differences between treatments, there did not appear to be any significant effects due to reciprocation, organic loading rates, or planted vs. not planted.

Average pH values among treatments ranged from 6.3 to 7.3 and were influenced by reciprocation and to a much lesser extent planting regime (Figure 4). With reciprocation, average pH values were at least one pH unit greater than the non-reciprocating controls, while treatments with and without plants differed by 0.1 pH unit. Among the two reciprocating treatments (six vs. twelve cycles per day), there were only marginal differences. Among MRS loading rates (50, 100, 200, and 400 mg/L), there were no discernable differences in average pH values: 7.0 vs. 7.0 vs. 6.9 vs. 6.9, respectively.

Reciprocation was effective in striping dissolved CO_2 from the wastewater due to aeration induced by air-lifts and the fill and drain process. In this study, reducing CO2 concentrations via reciprocation affected the carbonate-bicarbonate alkalinity system leading to increases in pH. Average values for reciprocating treatments were all greater than 7.0, while non reciprocating treatments were near or below 6.5. Values of pH above 6.5 are often accompanied by precipitation of calcium salts including apatite and calcium carbonate (Boyd 1982).

EC, a measure of total dissolved salts, is measured in umhos/cm. In natural waters, EC ranges from 20 to 1,500 umhos/cm (Boyd 1982). Average EC values for treatments (Figure 5), ranged from 1087 to 1392, and revealed that neither MRS loading rates nor planting regimes influenced EC values. However, treatments with six or 12 reciprocation cycles per day had average EC values 14.4 and 17.4% less than overall treatment averages, respectively. Factors that may have influenced these reductions in EC include 1) nutrient removal (TAN and nitrate-N), via microbial-based oxidation /reduction reactions and

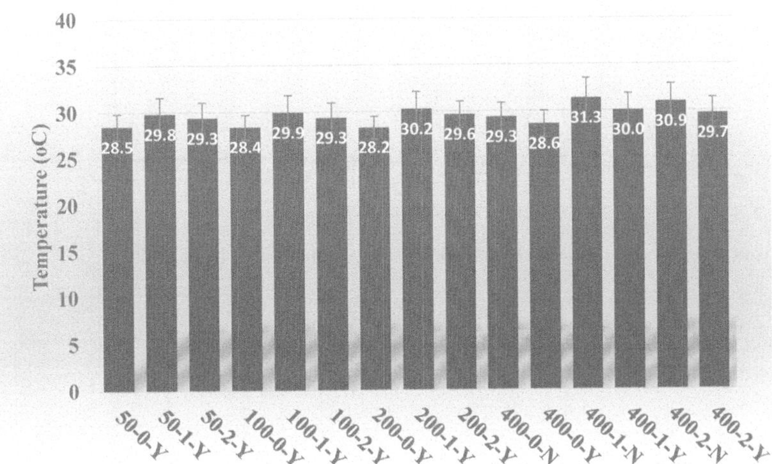

Figure 3 | Average temperatures (+/- one standard deviation) as a function of treatment.

Average pH values as a function of treatment

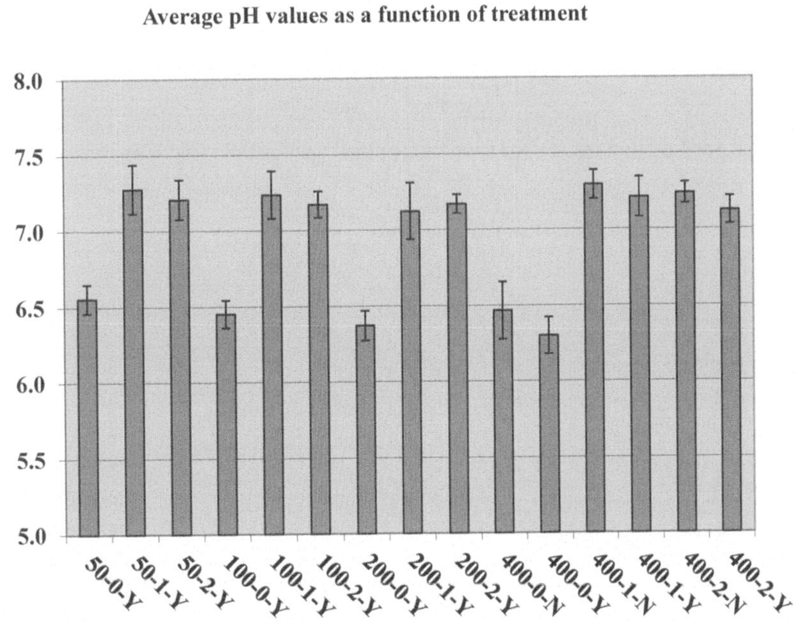

Figure 4 | Average pH values (+/− one standard deviation) as a function of treatment.

plant uptake 2) pH induced precipitation of salts such as calcium carbonate and apatite. Figure 6 illustrates the strong relationship between EC and pH for all 15 microcosm treatments with and without reciprocation.

Notice that when pH was high, conductivity was low and vice versa. This relationship was induced primarily by reciprocation and its effective stripping of dissolved CO_2.

DO dynamics

Within hours of starting the study, D concentrations in all treatments were reduced by 2.5–7.0 mg/L, indicating a robust microbial population and high community respiration rates (Figure 7). Irrespective of treatment, D concentrations reached their lowest point within the first 12 hours. After 12 hours, DO-related treatment dynamics rapidly diverged with oxygen

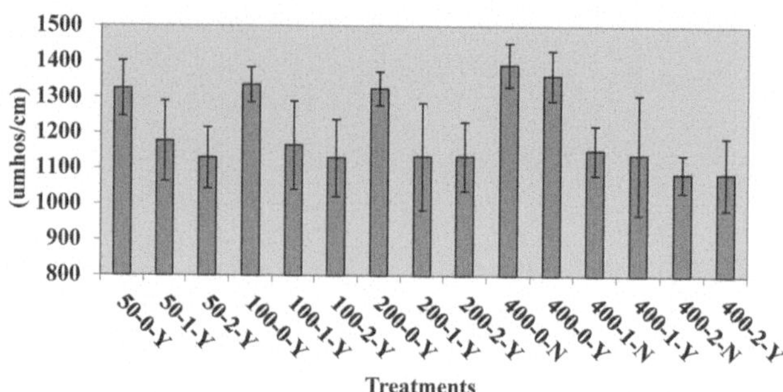

Figure 5 | Average conductivity values (+/− one standard deviation) as a function of treatment.

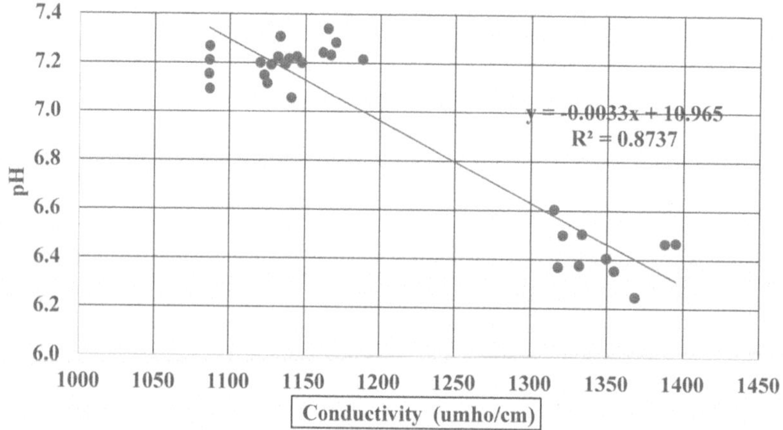

Figure 6 | Relationship between pH and EC (umhos/cm), as a function of treatment. Paired values are based on average values for replicates among the 15 treatments.

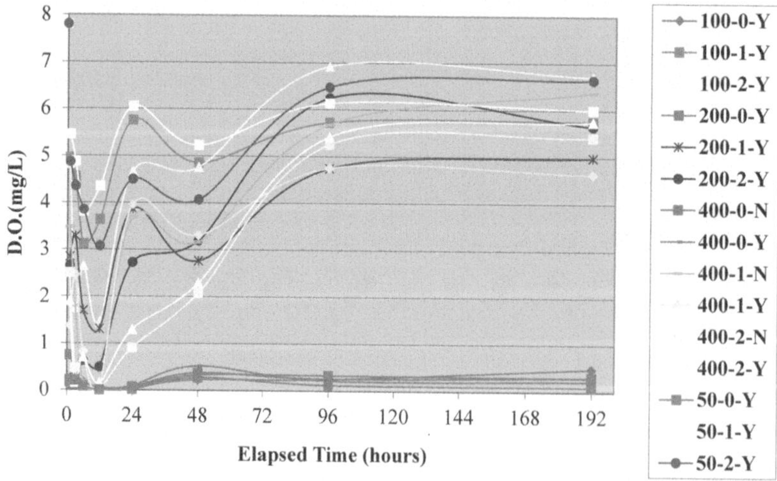

Figure 7 | D.O. (mg/L), concentration dynamics as a function of treatment and time.

concentrations increasing for all reciprocating treatments, while D concentrations for non-reciprocating treatments remained at <0.5 mg/L for the duration of the eight-day study.

Treatment D averages clearly revealed the effect of reciprocation vs. no reciprocation, and the lesser impact of MRS loading rates (Figure 8). While the reciprocating process resulted in higher average D concentrations, the differences among the

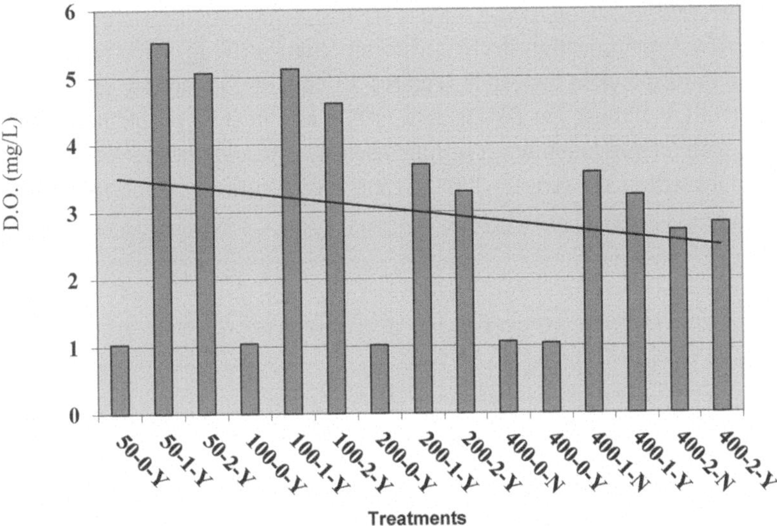

Figure 8 | D concentrations averaged by treatment over the 192-hour trial. The black trendline illustrates the relationship between MRS loading rates and average D concentrations.

reciprocating treatments (12 vs. six cycles/day) were marginal, amounting to less than 0.5 mg/L. With respect to MRS loading rates, average D concentrations were stepwise as expected, with 50 MRS > 100 > 200 > 400 (see trendline). The impact of plants vs. no plants on D concentration was marginal, amounting to less than 0.5 mg/L (see MRS 400 treatments).

Figure 9 illustrates average D concentrations at time zero vs. concentrations after 96 and 192 hours of treatment. This figure reveals marginal D recovery for non-reciprocating treatments even after 192 hours (8 days). Furthermore, among reciprocating treatments, most of the increases in oxygen concentration occurred with the first 96 hours. None of the treatments recovered to oxygen saturation during the study indicating significant residual community respiration.

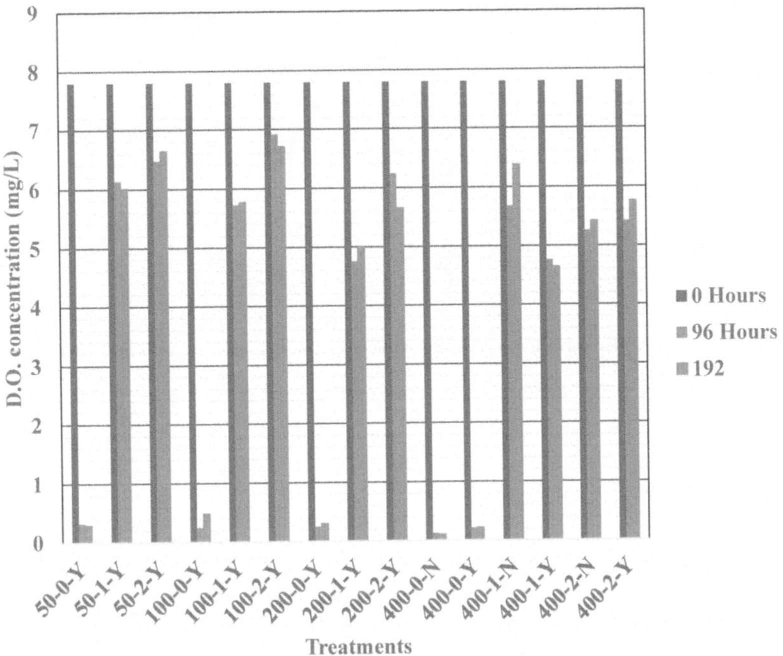

Figure 9 | Average D concentrations (mg/L) at start of experiment (0 hours) vs. treatment averages after 96 and 192 hours. Notice the low D concentrations for non-reciprocating treatments. Among reciprocating treatments, 95% of increases in D concentrations occurred within the first 96 hours.

Oxidation-reduction potential (ORP)

ORP values for treatments without reciprocation decreased to less than −200 to −300 mV within 24 hours and remained below −100 mv for the duration of the eight-day study (Figures 10 and 11). These low redox values are near optimum for production of methane, a potent greenhouse gas (Wang *et al.* 1993) and may be problematic for passive SSFW (Brix *et al.* 2001). In contrast, treatments with reciprocation never dropped below +80, and beyond 48 hours all reciprocating treatments maintained average redox values within the range +200 to +300 mV (Figure 10). Among the fifteen treatments evaluated,

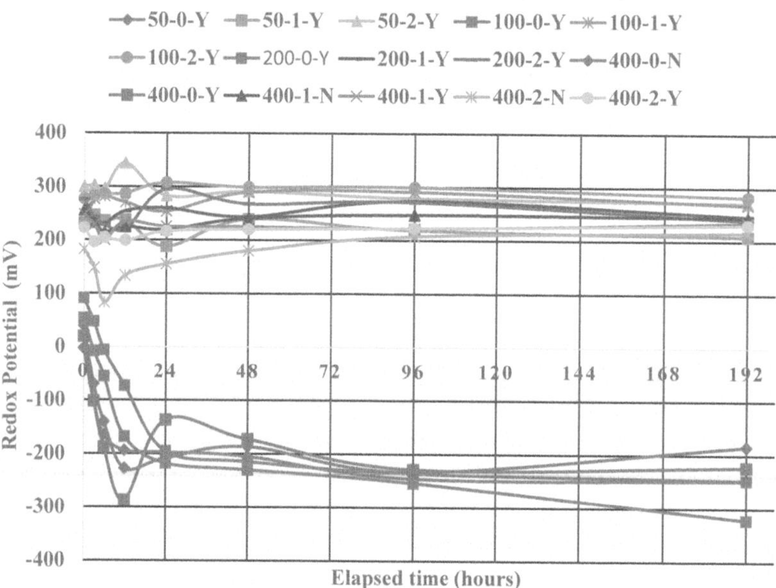

Figure 10 | Redox potential (mV), as a function of elapsed time and treatment. Data illustrated is based on averages of two replicates per treatment. Red lines denote treatments without reciprocation.

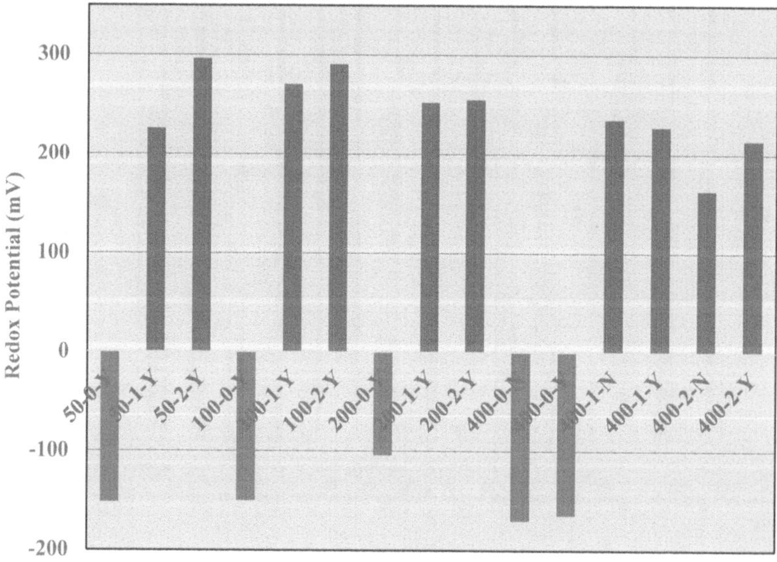

Figure 11 | Average redox potential (mV), as a function of treatment. Non-reciprocating treatments had consistently low values in the range conducive to methane production, while reciprocating treatments had values in the range needed for aerobic wastewater treatment, irrespective of COD concentrations or planting regimes. Red bars denote negative redox values for treatments without reciprocation.

there were significant correlations between D (mg/L) and redox potential (mV), and correlations between redox potential and pH (Figures 12 and 13, respectively).

COD

Batch-loaded COD concentrations for MRS 50, 100, 200 and 400 treatments were equivalent to 2.8, 5.7, 11.4, and 22.8 (g/m^2) respectively. Changes in COD were rapid and concentration dependent (Figure 14), with 46–95% reductions within the first 24 hours. COD removal among treatments with reciprocation (six vs. 12 cycles/day), averaged only 2 to 3% different. Thus, within the COD concentrations evaluated, there was a highly significant advantage to reciprocation, but no clear advantage to increasing the reciprocation regime from six to 12 cycles/day. This is an important finding, as pumping costs associated with reciprocation can be reduced by half by adopting the six cycles per day treatments.

COD removal as a function of planting regime (Y vs. N) was only evaluated at the 400 mg/L COD concentration. Results showed that COD removal was marginally more rapid for treatments without plants, which is consistent with the findings of Zhu & Sikora (1995), who found that canary grass roots, *P. arundinacea*, consistently released high concentrations of organic carbon. As will be seen later, this production of organic carbon can be important for enhancing removal of nitrate via microbial denitrification.

Figure 15 summarizes residual COD concentrations (mg/L) after 192 hours of treatment and final treatment averages expressed as percent removal. Residual COD concentrations ranged from 14 to 56 mg/L, with higher values associated

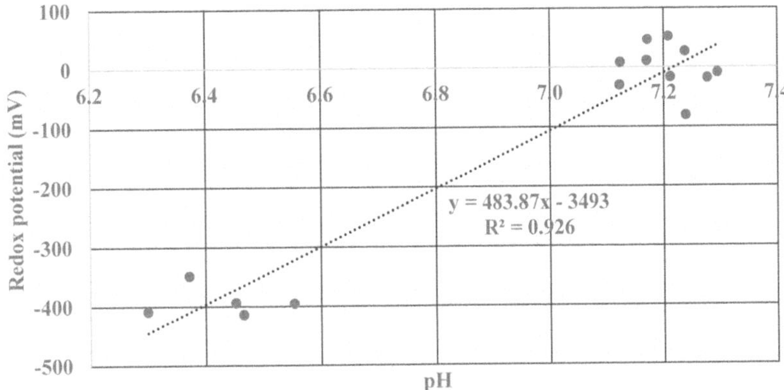

Figure 12 | Relationship between redox potential (mV) and pH as a function of treatment. Paired values are based on average values for replicates among the 15 treatments.

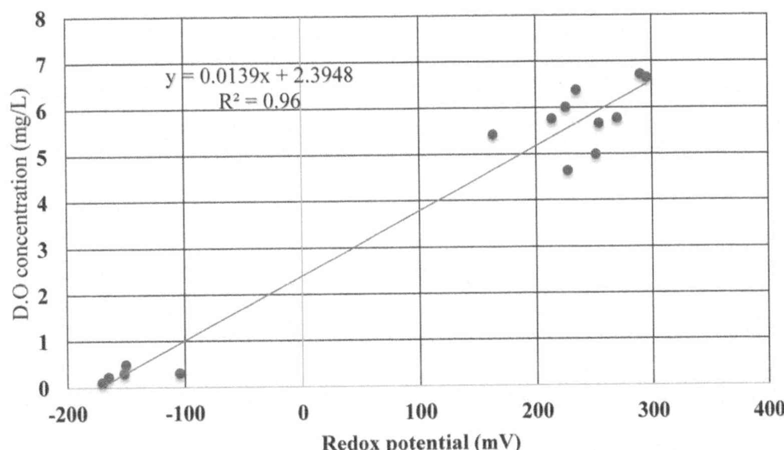

Figure 13 | Relationship between D (mg/L), and redox potential (mV) as a function of treatment. Paired values are based on average values for replicates among the 15 treatments.

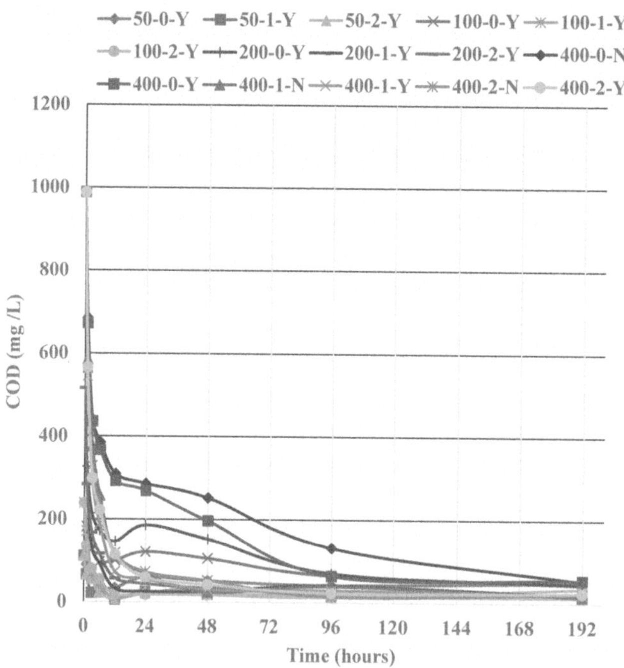

Figure 14 | COD dynamics as a function of MRS loading rate (mg/L) (50,100, 200, 400), reciprocation regime (0 vs. six vs.12), and planting regime (Yes vs. No; with and without *P. arundinacea*). Note the rapid removal rates during the first six to 12 hours, especially for reciprocating treatments.

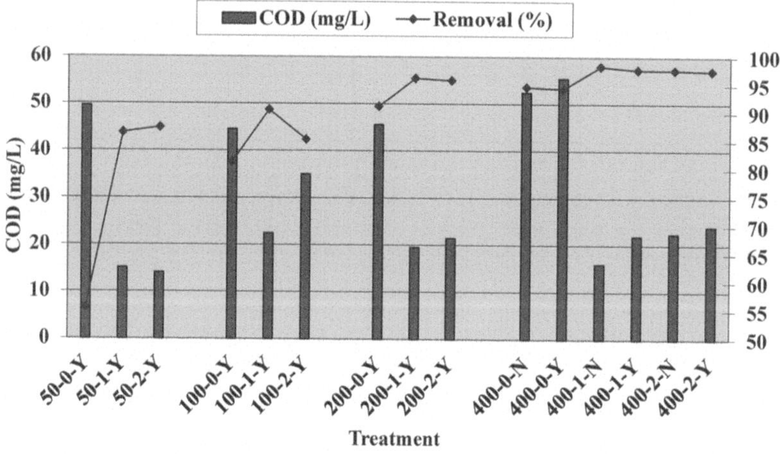

Figure 15 | Residual COD concentrations (mg/L), and removal (%), after 196 hours as a function of treatment.

with non-reciprocating treatments. Removal of COD expressed as a percent, ranged from 55 to 98%, with the higher removal rates associated with treatments receiving higher COD concentrations. This is consistent with first-order removal rates and has implications for placement of decentralized systems. Placement of decentralized reciprocating systems closer to the concentrated wastewater source provides higher removal rates and thus more economical treatment. The data also reveals that most of the COD removal in reciprocating systems occurred within the first 48 hours. This finding implies that a two-day retention time for reciprocating systems may be most economical and near optimum for COD removal within the COD concentrations tested (111, 238, 516, 988 mg/L).

Values of k (first-order removal rate coefficients) ranged from 0.10 to 0.55, and revealed progressively higher K values with increasing COD concentrations (Figure 16). Values of k were also higher for treatments with reciprocation. Note that in the

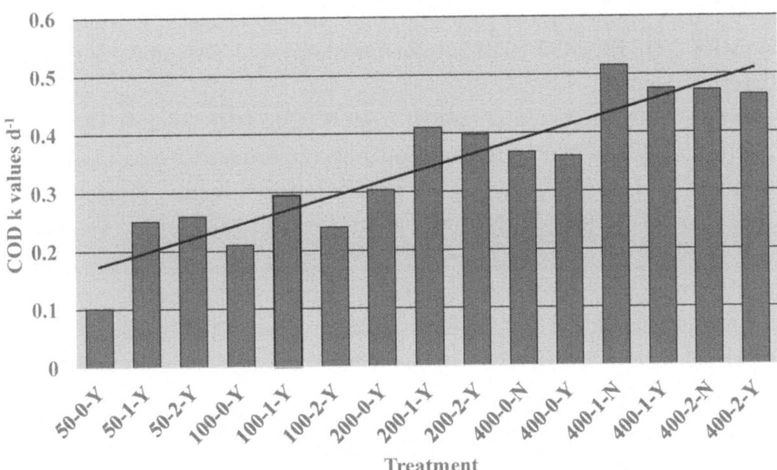

Figure 16 | COD k values d^{-1} as a function of treatment based on first-order removal rate kinetics: according to ln (C$_0$/C) = kt, where C and C$_0$ represent concentrations at time t and the starting concentration, respectively. Trendline indicates higher removal rate coefficients with increases in organic loading rate.

400 MRS treatments (988 mg/L C OD), rates of removal were marginally higher for treatments without *P. arundinacea*. For treatments without reciprocation and low D.O. concentrations, it is surmised that significant metabolism occurred under anoxic/anaerobic conditions with CO$_2$ and methane as end-products. Methanogenesis is a more efficient process for treating recalcitrant organic matter and produces less sludge than aerobic metabolism. However, anaerobic metabolism produces methane, a potent greenhouse gas and can require longer retention times to completely degrade recalcitrant organic compounds (Zitomer & Speece 1993).

Total ammonia nitrogen (TAN)

TAN batch loading to each of the 15 treatments was equivalent to 5.7 g/m^2 with an initial starting concentration of 100 mg/L. Removal of TAN was rapid for all treatments within the first six hours, with reciprocating treatments exhibiting significantly faster removal as compared to non-reciprocating treatments (Figure 17). Removal rates, expressed as percentages, are illustrated for each treatment at 48 and 96 hours (Figure 19). TAN removal for reciprocating treatments ranged from 88–99% at 48 hours and greater than 99% at 96 hours. In contrast, non-reciprocating treatments removed from 52–64% at 48

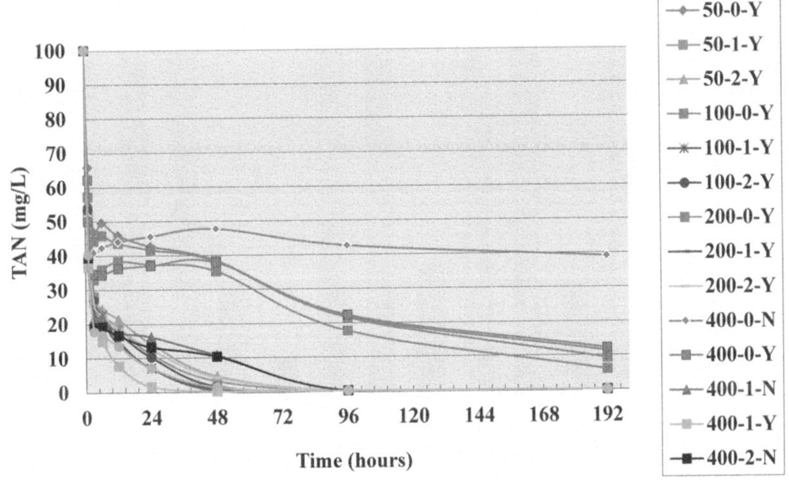

Figure 17 | TAN dynamics as a function of MRS treatment (50,100, 200, 400), reciprocation regime (0 vs. six vs. 12 cycles), and planting regime (Yes vs. No). Red highlighted data references removal dynamics for non-reciprocating treatments.

hours and from 57–82% at 96 hours. The impact of canary grass, *P. arundinacea,* in non-reciprocating treatments receiving the highest organic loading (MRS 400), revealed 10 to 35% better removal of TAN at 48 and 96 hours respectively, as compared to the non-planted controls (Figure 18).

Values of k (first-order removal rate coefficients, ranged from 0.12 to 0.96 (Figure 19). Removal rate coefficients for reciprocating treatments were 2.6 to 8.0 times greater than non-reciprocating treatments. Organic loading rates did not appear to significantly affect TAN k-values among reciprocating treatments. However, in non-reciprocating treatments at the highest loading rate (MRS 400), TAN removal was greater for planted vs. not planted; k = 0.26 vs 0.12, respectively.

It is proposed that greater TAN removal rates among reciprocating treatments was due primarily to the repeated exposure of the microbial biofilms to atmospheric oxygen and the ensuing positive impact of oxygen availability for nitrifying bacteria. Furthermore, reciprocation supplied sufficient oxygen to meet both the oxygen requirements for removal of COD and

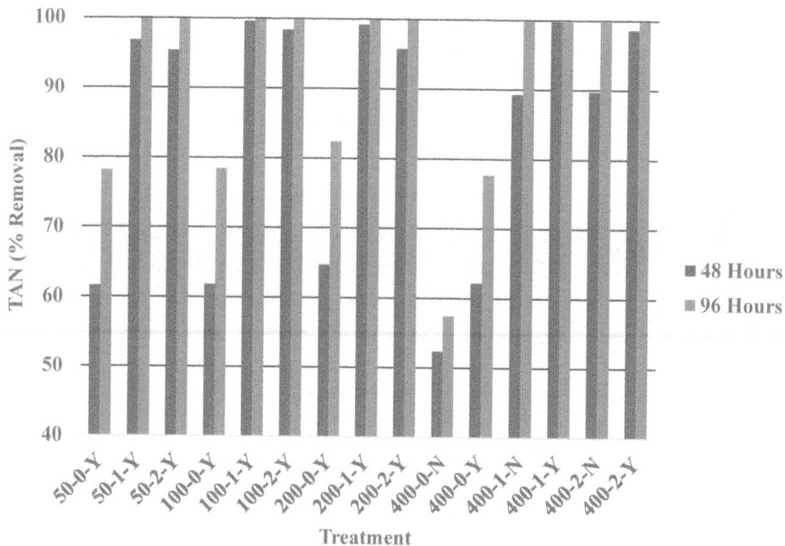

Figure 18 | TAN percent removal at 48 and 96 hours as a function of treatment.

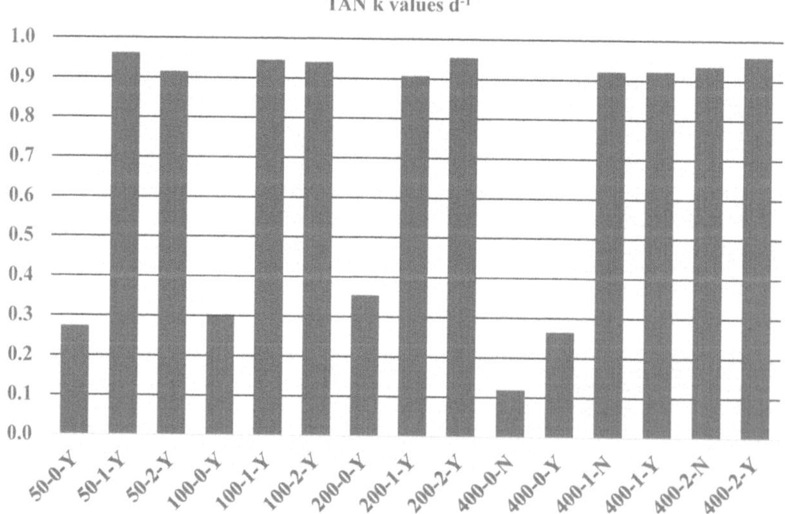

Figure 19 | TAN k values d^{-1} as a function of treatment. Notice the consistently higher k values for reciprocating treatments vs. non-reciprocating treatments.

removal of TAN. Uptake of ammonium by plants was also clear (Figure 18), as revealed by the direct comparison of TAN removal rates in treatments in the high organic loading treatments (400 MRS). TAN removal rates for treatments with *P. arundinacea* were 18 to 35% higher than for treatments without plants.

Nitrate nitrogen

There was no nitrate added to the initial synthetic wastewater solutions, and thus any nitrate detected in the treatment systems was due to *in situ* nitrification. Denitrification is defined as the 'process in which nitrate is converted to dinitrogen gas (N_2), via intermediates nitrite, nitric oxide, and nitrous oxide' (Vymazal & Kropfelova 2008). Denitrification is a microbial driven process and occurs in anoxic and anaerobic environments and requires a bioavailable dissolved organic carbon source (Garcia *et al.* 2010).

In the present study, total nitrogen dynamics was influenced by nitrification and denitrification, reciprocation and the presence or absence of canary grass, *P. arundinacea*, a plant known for its ability to reduce rhizosphere redox potential (Steinberg & Coonrod 1994; Zhu & Sikora 1995). Canary grass is a valuable wetland plant as it can uptake ammonium and nitrate as nutrient sources and supply a root-exuded organic carbon source, which supplements the carbon needed for denitrification (Zhu & Sikora 1995). Figure 20 illustrates that in 13 of the 15 treatments evaluated (all with canary grass), nitrate was only detected at low concentrations (<4.0 mg/L) during the first 48 hours of the study and were less than one mg/L after 96 hours of treatment. This shows that while nitrification was ongoing in these treatments at a high rate (Figure 17), denitrification, microbial immobilization, and plant uptake were removing nitrate as fast as it was being produced, and thus there was no significant buildup of nitrate. This is also consistent with the findings of Sikora *et al.* (1995a). Conversely, treatments with reciprocation (six and 12 cycles per day), high organic inputs (400 MRS), and no canary grass showed nitrate concentrations of 17 and 27 mg/L after 96 hours of operation (Figure 20). As noted earlier, microbial denitrification requires anoxic conditions and a labile organic carbon source. While reciprocation was effective at increasing oxygen supply and rapidly reducing COD, there was no canary grass to supplement the organic carbon, and therefore denitrification was carbon limited. If denitrification is needed in reciprocating systems, there will be a need for carbon supplementation or use of plants that supply a labile carbon source (Zhu & Sikora, 1995).

Phosphorus

Phosphorus batch loading to each of the 15 treatments was equivalent to 2.85 g/m^2 with an initial starting concentration of 50 mg/L. Removal of P was rapid during the first 12 hours and then tended to stabilize. The rapid removal was influenced by

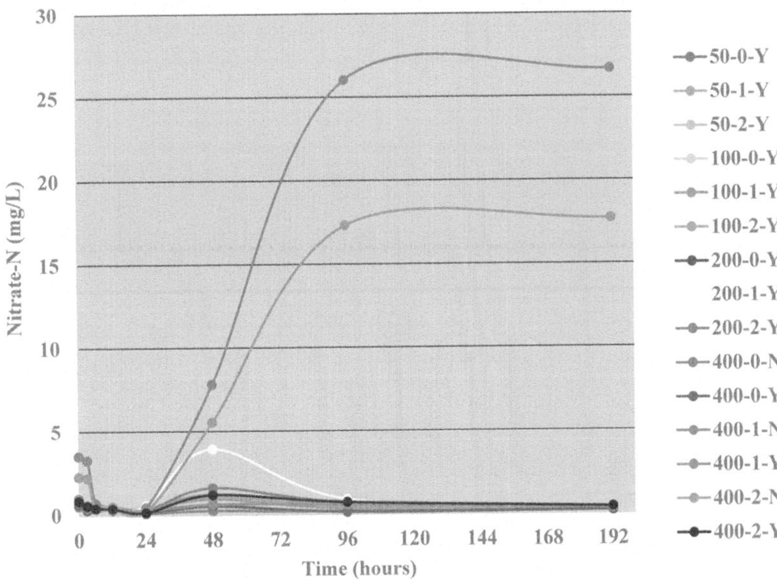

Figure 20 | Nitrate dynamics as a function of treatment and time. Notice that only treatments without canary grass, *P. arundinacea,* had significant nitrate accumulation.

sorption of P to the rock substrate as a function of exchange sites (Sikora *et al.* 1995a; Zhu *et al.* 1997). After eight days, average removal rates ranged from 92 to 100% (Figure 21). Figure 22 illustrates treatment averages and trendlines at 24 and 192 hours. Data shows that after 24 hours, treatments with reciprocation removed significantly more P than treatments without reciprocation. Also, at the highest COD concentration (400 MRS), P removal was greater with plants than without plants.

Shi *et al.* (2017), evaluated P-removal in intermittently aerated wetlands and found superior plant growth and P removal as compared to non-aerated controls. Their research also evaluated several different substrates and found fly-ash brick to be sig-

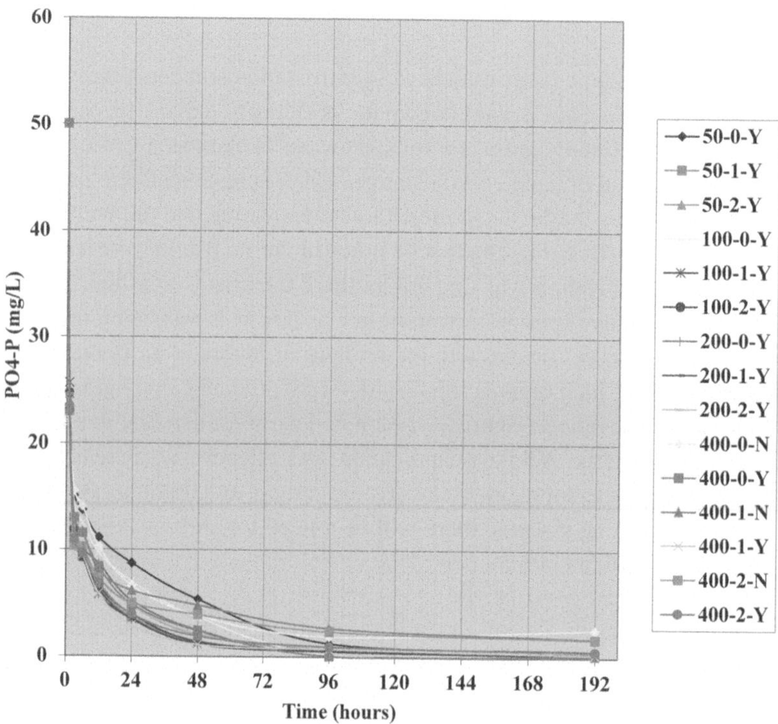

Figure 21 | Phosphorus dynamics as a function of treatment and time.

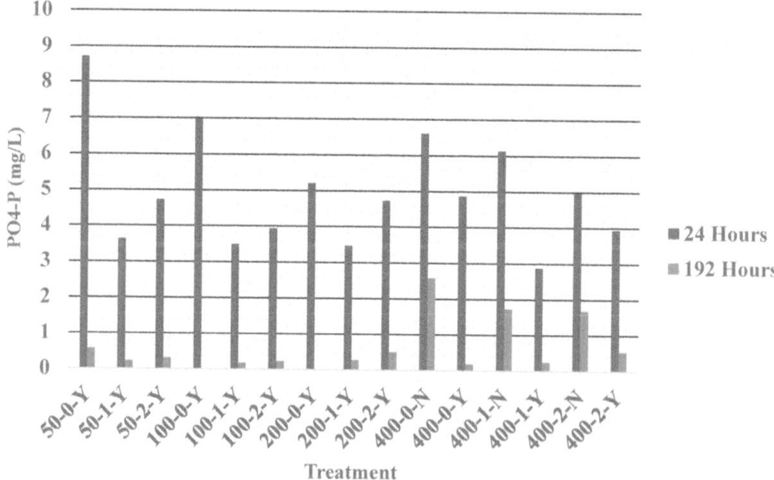

Figure 22 | Phosphorus dynamics as a function of treatment and time. Data illustrates average treatment concentrations after 24 and 192 hours of treatment. Starting concentration was 50 mg/L.

nificantly better than gravel or red clay brick. Ilyas & Masih (2018), found tidal flow wetlands (like reciprocating wetlands) to be superior to other wetland types for P removal and reported removal rates of 88% and 2.6 g/m^2/day. They attributed the superior performance to 'a positive effect of tidal flow on rejuvenating the wetland with fresh air, thus enhancing dissolved oxygen (DO) in the system, and augmenting phosphorus precipitation and adsorption to the substrate.'

While research results reported above are encouraging, sustained phosphorus removal in substrate based SSFW will be difficult to achieve, because removal dynamics in gravel-based systems are controlled primarily by adsorption of P to limited sorption sites on the substrates surface. Once these sites are saturated, P removal will be significantly diminished. While P uptake by plants and microbes can be significant, they are seasonal and not sustainable in the long term unless plant and microbial biomass are harvested and removed from the wetland site (Vymazal & Kropfelova 2008).

ET

ET is the joint loss of water from the landscape due to evaporation from surface substrates and transpiration, the loss of water through vegetation. ET is influenced by plant biomass, prevailing wind speed, and seasonal variations in temperature and humidity (Vymazal & Kropfelova 2008).

Figure 23 illustrates ET rates (mm/m^2/day) for 15 treatments during an eight-day batch-loaded treatability study. The trendline indicates that the influence of MRS loading rate was insignificant, while the influences of plants and reciprocation were additive and significant. Treatments with plants but no reciprocation averaged 2.7 mm/m^2/day, while treatments with plants and reciprocation (six and 12 cycles/day), averaged 8.5 and 10.8 mm/m^2/day, respectively.

Data for the MRS 400 series illustrates the influence of two interacting treatment variables: with and without reciprocation, and with and without plants. This data reveals a strong positive interaction for planting and reciprocation regimes. Average ET rates (mm/day) were low for treatments without plants or reciprocation (1.2) and moderately low for treatments with no plants but with reciprocation (4.0). However, with both plants and reciprocation, average ET was equivalent to 10.9 mm/m^2/day).

The ET values reported in this study are within the range of values reported in the literature (Vymazal & Kropfelova 2008; Milani *et al.* 2019). While the data in the present studies were derived from a short-term treatability study with low plant biomass, the among-treatment values reveal the positive interaction of reciprocation and planting regimes on ET and water budgets.

Microcosm utility and scaling to field applications

The research team designed and fabricated the 30-tank microcosm system on-site using off-the-shelf items. During this eight-day study and four later studies, the microcosm system ran admirably with few problems. Use of solenoid valves, digital

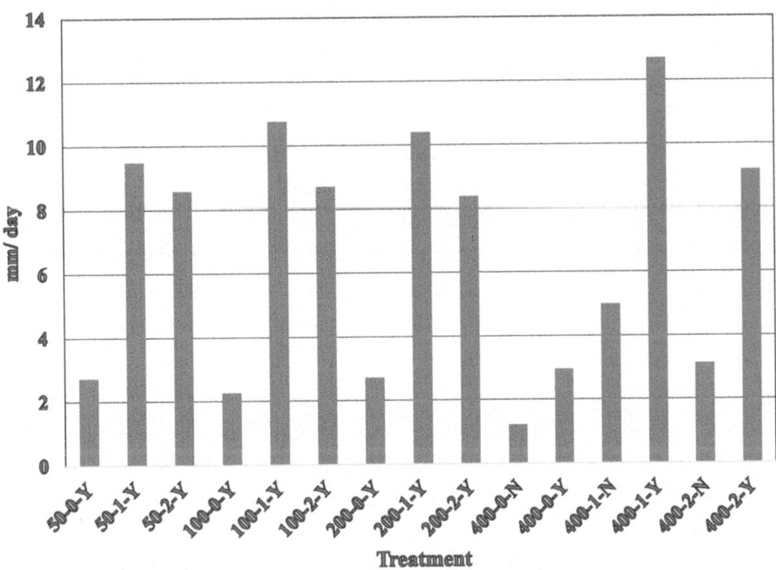

Figure 23 | Average daily evapotranspiration rates (mm/day) as a function of treatment.

timers, regenerative blowers, and one-way aquaria valves allowed air-lift operations to distribute discrete and uniform volumes of water between treatment cells on a timed basis. Therefore, between tank experimental error was well controlled as corroborated by low coefficients of variation (CVs), for measured variables. For example, the CV for TAN over 120 determinations was 12.4%.

This microcosm system was cost-effective with respect to materials and operating costs and enabled complex experimental designs in which main effects and interactions could be quantified. Future studies should be designed to further evaluate (1) experimental substrates and substrate combinations, (2) aquatic and terrestrial plants for applications of phytoremediation with- and without reciprocation (3) optimized cycle times for high strength anaerobic lagoon wastewater, and (4), engineered underdrain designs for long term storage and treatment of bio-solids. Because reciprocation aids movement of sluffed biosolids to engineered underdrains, this may enhance anaerobic degradation of solids and mitigate clogging issues and extend the life of the treatment system.

The recurrent fill and drain process provided mixing of the wastewater and passive aeration such that nutrients and oxygen were distributed uniformly to the biofilms and to the plant roots in the rhizosphere. It is envisioned that reciprocation and attendant mixing will diminish, if not eliminate, short circuiting which is a problem in conventional surface-flow and subsurface-flow constructed wetlands (Dierberg *et al.* 2005).

With reciprocation, atmospheric oxygen was available during the drain phase. Nivala *et al.* (2013) monitored oxygen consumption rates in fifteen different pilot-scale wetland treatment systems and reported maximum oxygen consumption of 89.3 g/m^3/day for a 2-cell reciprocating system treating municipal wastewater. This was the highest rate among the fifteen different treatment systems examined and testifies to the utility of the reciprocating process to provide a near-inexhaustible supply of atmospheric oxygen for aerobic wastewater treatment processes. Conversely, because of the anoxic environment that develops during the fill phase of reciprocation, denitrification, and TN removal were also significantly greater than the other 14 pilot-scale systems being evaluated. Furthermore, in the present study it was revealed that reciprocation allowed supersaturated metabolic gases, such as CO_2, to escape such that pH values were increased by one full unit as compared to non-reciprocating controls. This may have utility for enhancing calcite precipitation and removal of phosphorus.

The exceptionally high removal rates for COD, TAN, and P during the first 3 to 12 hours was due to a combination of factors including high initial concentrations, microbial immobilization, plant uptake of nutrients, and the adsorption of TAN COD, and P to rock substrates due to cation and metal binding exchange sites. Sikora *et al.* (1995a, 1995b), provide relevant and detailed interpretations and discussions of cation exchange dynamics for gravel-based constructed wetland systems and notes that with respect to ammonium, adsorption can be rapidly reversible due to the dynamic equilibrium with nitrification. It should also be noted that the rapid removal of DO during the first 12 hours is emblematic of high microbial community respiration rates associated with COD removal and nitrification (Nivala *et al.* 2013).

It is surmised that having an aerobic rhizosphere in reciprocating systems will allow culture of select terrestrial plant species for purposes of phytoremediation, including flowers, woody ornamentals, tree saplings, and fodder crops. It should be noted that these plants have economic and aesthetic values beyond their wastewater treatment functions.

While these short-term batch-loaded microcosm treatability studies provided significant treatment of synthetic wastewater, later collaborative and third-party pilot-scale and commercial-scale demonstrations have corroborated the sustainability, scalability, and energy efficiency (Austin & Nivala 2009), of deep-cell reciprocating systems for treating a diversity of domestic, industrial, and agricultural wastewaters. Examples include airport deicers (Arendt *et al.* 2002), acid mine drainage (Sikora *et al.* 1995b), swine lagoon wastewater (Behrends *et al.* 2004), dairy lagoon wastewater (Choperena 2010; Henneman 2011), pretreated high-density aquaculture wastewater (Behrends *et al.* 1999, 2007) and domestic wastewater (Nivala *et al.* 2019). Reciprocating wetlands have also been evaluated to control malodors (swine) and greenhouse gas emissions (dairy) and found to have significant mitigation capabilities (Schiffman *et al.* 2008; Henneman 2011).

Conceptually, reciprocating wetlands may have even greater economic value for wastewater treatment in the tropics and at high altitudes, where oxygen solubility is reduced due to elevation and wastewater temperatures respectively (Baquero-Rodrigues *et al.* 2022). It should be noted that reciprocation requires electricity for rapidly moving water (fill and drain), among paired treatment cells. However electrical costs can be significantly reduced by using auto-siphons and axial flow pumps that are designed for low head-high volume applications Yoo & Boyd (1994). Clogging of wetland gravel substrates is also a potential problem for all gravel-based systems, including reciprocating wetlands (Behrends *et al.* 2006). Regarding control and management of clogging, Nivala *et al.* (2012) concluded 'the best approach moving forward will be to minimize the negative effects of clogging through improved pretreatment, better systems design, and new operation and management practices.

CONCLUSION

This short term eight-day microcosm study quantified the impacts of reciprocation with and without plants (canary grass, *P. arundinacea*) on the treatment dynamics of COD, TAN, nitrate, phosphorus, and ET. Reciprocation induced passive aeration and the development of recurrent aerobic and anoxic environments six to 12 times per day. During the drain phase, the microbial biofilms were exposed to high atmospheric oxygen concentrations (21%), which promoted rapid microbial oxidation of ammonium and COD and removal of phosphorus. During the fill phase, the biofilms were covered by oxic-anoxic water, which enhanced denitrification and the removal of nitrate. The combination of reciprocation with *P. arundinacea* significantly improved biological denitrification, ET, and other wastewater treatment dynamics.

Among treatments with six and 12 reciprocation cycles per day there were significant improvements with respect to removal of COD, ammonium, nitrate, and phosphorus. Treatments with reciprocation and *P. arundinacea* removed 81–98% of COD, greater than 95% of ammonium and nitrate, and greater than 98% of phosphorus. Treatments with both reciprocation and plants (*P. arundinacea*), revealed a strong positive interaction with respect to ET, with treatment averages ranging from 8.3 to 12.7 mm/day.

ACKNOWLEDGEMENTS

This work was funded and supported by the Tennessee Valley Authority's Environmental Research Center (TVA), and by grants from the U.S. Environmental Protection Agency (EPA).

DATA AVAILABILITY STATEMENT

All relevant data are included in the paper or its Supplementary Information.

CONFLICT OF INTEREST STATEMENT

The authors declare there is no conflict.

REFERENCES

Arendt, T., Ervin, M. & Florea, D. 2002 Reciprocating subsurface flow treatment system keeps airport out of the deepfreeze. In *Water Environment Federation Technical Exhibition and Conference*, Sept. 28 - Oct. 2. McCormick Place Convention Center Chicago IL.

Armstrong, W., Cousins, D., Armstrong, J., Turner, D. & Beckett, P. 2000 Oxygen distribution in wetland plant roots and permeability barriers to gas-exchange with the rhizosphere: a microelectrode and modelling study with Phragmites australis. *Annals of Botany* **86**, 687–703.

Austin, D. & Nivala, J. 2009 Energy requirements for nitrification and biological nitrogen removal in engineered wetlands. *Ecological Engineering* **35**, 184–192.

Baquero-Rodrigues, G., Martinez, S., Acuna, J., Nolasco, D. & Rosso, D. 2022 How elevation dictates technology selection in biological wastewater treatment. *Journal of Environmental Management* **307**, 114588.

Behrends, L. 1999 *United States Patent 5,863433*. United States Patent Office, Washington DC.

Behrends, L., Sikora, F., Coonrod, H., Bailey, E. & Bulls, M. 1996 Reciprocating subsurface-flow constructed wetlands for removing ammonia, nitrate, and chemical oxygen demand: potential for treating domestic, industrial, and agricultural wastewaters. In: *Proceedings of WEFTEC_ '96, the 69th Annual Conference and Exposition of the Water Environment Federation, Vol. 5, Part II 9610005*. Water Environment Federation, Alexandria Virginia.

Behrends, L., Houke, L., Bailey, E. & Brown, D. 1999 Reciprocating subsurface-flow constructed wetlands for treating high-strength aquaculture wastewater. In: *Wetlands and Remediation: An International Conference* (Means, J. L. & Hinchee, R. E. eds.). Salt Lake City, Utah, pp. 317–325

Behrends, L., Bailey, E., Ellison, W., Houke, L., Jansen, P., Shea, C., Smith, S. & Yost, T. 2004 Reciprocating constructed wetlands for treating high strength anaerobic lagoon wastewater. In *American Society of Agricultural Engineers, Animal, Agricultural and Food Processing Waste IX, Conference Proceedings*.

Behrends, L., Bailey, E., Houke, L., Jansen, P. & Smith, S. 2006 Non-invasive methods for treating and removing sludge from subsurface-flow constructed wetlands II. In: *Proceedings of the 10th International Conference on Wetland Systems for Water Pollution Control*, 23–29 September 2006. Ministerio de Ambiente, do Ordenamento do Territori e do Desenvolvimento Regional and IWA, Lisbon Portugal, pp. 1271–1281.

Behrends, L., Houke, L., Bailey, E., Jansen, P. & Smith, S. 2007 Integrated constructed wetland systems: design, operation, and performance of low-cost decentralized wastewater treatment systems. *Water Science & Technology* **55** (7), 155–161.

Boyd, C. E. 1982 *Water Quality Management for Pond Fish Culture*. Elsevier Scientific Publishing Company, New York, p. 318.

Brix, H. & Schierup, H.-H. 1990 Soil oxygenation in constructed reed beds: the role of macrophyte and soil-atmosphere interface oxygen transport. In: *Constructed Wetlands in Water Pollution Control*, Cooper P.F., and Findlater B.C., (eds.), Pegamon Press, Oxford, United Kingdom, pp. 53–66.

Brix, H., Sorrell, B. & Lorenzena, B. 2001 Are Phragmites-dominated wetlands a net source or net sink of greenhouse gases? *Aquatic Botany* **69**, 313–324.

Choperena, J. 2010 *Demonstration and Evaluation of A Reciprocating Biofilter for Dairy Lagoon Wastewater*. Final report: U.S. EPA Region 9 Funded Grant ID#: 9694001.

Cooper, P. & Findlater, B. eds., 1990 *Constructed Wetlands in Water Pollution Control*. Pergamon Press, Oxford.

Crites and Tchobanoglous 1998 *Small and Decentralized Wastewater Management Systems*. McGraw Hill, New York, USA.

Dierberg, F., Juston, J., DeBusk, T., Pietro, K. & Gu, B. 2005 Relationship between hydraulic efficiency and phosphorus removal in a submerged aquatic vegetation-dominated treatment wetland. *Ecological Engineering* **25**, 9–23.

García, J., Aguirre, P., Barragán, J., Mujeriego, R., Matamoros, V. & Bayona, J. M. 2005 Effect of key design parameters on the efficiency of horizontal subsurface flow constructed wetlands. *Ecological Engineering* **25** (2005), 405–418.

Garcia, J., Diederik, P. L., Rousseau, L., Morato, J., Lesage, E., Matamoros, V. & Bayona, J. 2010 Contaminant removal processes in subsurface-flow constructed wetlands: a review. *Critical Reviews in Environmental Science and Technology* **40** (7), 561–661.

Green, M., Friedler, E., Ruskol, Y. & Safrai, I. 1997 Investigation of alternative method for nitrification in constructed wetlands. *Water Science and Technology* **35** (5), 63–70.

Henneman, S. 2011 *Water and Air Quality Performance of A ReCiprocating Biofilter Treating Dairy Wastewater*. Master's Thesis, California Polytechnic State University San Luis Obispo.

Ilyas, H. & Masih, I. 2018 The effects of different aeration strategies on the performance of constructed wetlands for phosphorus removal. *Environmental Science and Pollution Research International* **6**, 5318–5335.

Milani, M., Marzo, A., Attilio Toscano, A., Consoli, S., Giuseppe, G., Cirelli, L., Ventura, D. & Barbagallo, S. 2019 Evapotranspiration from horizontal subsurface flow constructed wetlands planted with different perennial plant species. *Water* **11** (10), 2159.

Nivala, J., Knowles, P., Dotro, G., Garcia, J. & Wallace, S. 2012 Clogging in subsurface-flow treatment wetlands: measuring, modeling, and management. *Water Research* **46** (2012), 1625–1640.

Nivala, J., Wallace, S., Headley, T., Kassa, K., Brix, H., Afferden, M. & Muller, R. 2013 Oxygen transfer and consumption in subsurface flow treatment wetlands. *Ecological Engineering* **61** (Part B), 544–554.

Nivala, J., Boog, J., Headly, T., Aubron, T., Wallace, S., Brix, H., Mothes, S., Afferdvan, M. & Mueller, R. 2019 Side-by-side comparison of 15 pilot-scale conventional and intensified subsurface flow wetlands for treatment of domestic wastewater. *Science of The Total Environment* **658** (6), 1500–1513.

Rakocy, J. E. 2012 Aquaponics-Integrating Fish and Plant Culture. In: *Aquaculture Production Systems* (J. Tidwell, ed.). Wiley-Blackwell, Oxford, UK, pp. 344–386.

Schiffman, S., Graham, B. & Williams, C. 2008 Dispersion modeling to compare alternate technologies for odor remediation at swine facilities. *Journal of the Air & Waste Management Association* **58** (9), 1166–1176.

Shi, X., Fan, J., Zhang, J. & Shen, Y. 2017 Enhanced phosphorus removal in intermittently aerated constructed wetlands filled with various construction wastes. *Environ Sci Pollut Res* **24**, 22524–22534.

Sikora, F., Zhu, T., Behrends, L., Steinberg, S. & Coonrod, S. 1995a Ammonium removal in constructed wetlands with recirculating subsurface flow: removal rates and mechanisms. *Water Science and Technology* **32** (3), 193–202.

Sikora, F., Behrends, L. & Brodie, G. 1995b Manganese and trace metal removal in anaerobic and aerobic wetland environments. In *Proceedings American Power Conference, Chicago, IL (United States)*, 18–20 Apr 1995. Proceedings. Volume 57-II; PB: p. 914.

Steinberg, S. L. & Coonrod, H. S. 1994 Oxidation of the root zone by aquatic plants growing in gravel- nutrient solution culture. *Journal of Environmental Quality* **23**, 907–913.

Tyroller, L., Rousseau, D., Santa, S. & Garcia, J. 2010 Application of the gas tracer method for measuring oxygen transfer rates in subsurface flow constructed wetlands. *Water Research* **44** (14), 4217–4225.

Vymazal, J. & Kropfelova, L. 2008 *Wastewater Treatment in Constructed Wetlands with Horizontal Sub-Surface Flow*. Environmental Pollution 14. Springer Science and Business Media V, Berlin-Heidelberg, Germany.

Wang, Z., Delaunay, R., Masscheleyn, P. & Patrick Jr., W. 1993 Soil redox and pH effects on methane production in a flooded rice soil. *Soil Science Society of America Journal* **57 m** (2), 382–385.

Watson, J. & Danzig, A. 1993 Pilot-scale nitrification studies using vertical-flow and shallow horizontal-flow constructed wetlands cells. In: *Constructed Wetlands for Water Quality Improvement* (Gerald Moshiri, ed.). CRC Press Inc., Boca Raton, Florida, eBook ISBN9781003069997.

Wu, S., Dongxiao, Z., Austin, D., Dong, R. & Pang, C. 2011 Evaluation of a lab-scale tidal flow constructed wetland performance: oxygen transfer capacity organic matter and ammonium removal. *Ecological Engineering* **37** (11), 1789–1795.

Yoo, K. & Boyd, C. 1994 *Hydrology and Water Supply for Pond Aquaculture. Any AVI Book*. Chapman and Hall, New York, NY.

Zhu, T. & Sikora, F. 1995 Ammonium and nitrate removal in vegetated and unvegetated gravel bed microcosm wetlands. *Water Science and Technology* **32** (3), 219–228.
Zhu, T., Jenssen, P. D., Machlum, T. & Krogstad, T. 1997 Phosphorus sorption and chemical characteristics of lightweight aggregates (LWA): potential filter media in treatment wetlands. *Water Science and Technology* **35** (5), 103–108.
Zitomer, D. & Speece, R. 1993 Sequential environments for enhanced biotransformation of aqueous contaminants. *Environmental Science & Technology* **27**, 227–244.

First received 6 April 2022; accepted in revised form 1 June 2022. Available online 6 June 2022

doi: 10.2166/wst.2022.199

Performance of carbendazim removal using constructed wetlands for the Ethiopian floriculture industry

Stan Wehbe [a,*], Feleke Zewge[b], Yoshihiko Inagaki[c], Wolfram Sievert[d], N.T. Uday Kumar[e] and Akshay Deshpande[f]

[a] African Center of Excellence for Water Management, Water Science & Technology, Addis Ababa University, Addis Ababa, Ethiopia
[b] Department of Chemistry, Faculty of Science, Addis Ababa University, P.O. Box 1176, Addis Ababa, Ethiopia
[c] Department of Civil and Environmental Engineering, Waseda University, 3-4-1 Okubo, Shinjuku-city, Tokyo 169-8555, Japan
[d] Sievert Consult, Schuhstraße 15, 32657 Lemgo, Germany
[e] RAK Research and Innovation Centre, American University of Ras Al Khaimah (AURAK), Ras Al Khaimah, P.O. Box: 31208, United Arab Emirates
[f] Department of Research, Reed Bed Wastewater Treatment, Dubai, United Arab Emirates
*Corresponding author. E-mail: stanislas.vincent@aau.edu.et, stanwehbe@gmail.com

SW, 0000-0002-8334-4444

ABSTRACT

Carbendazim is a pesticide commonly used in Ethiopian flower farms and has harmful effects on aquatic, invertebrate, and mammalian life. Previous studies have explored ways to remedy carbendazim toxicity; however, the use of constructed wetland (CW) systems for carbendazim removal from farm water runoff has not been explored in depth. The primary aim of this study was to investigate the efficacy of a CW system for carbendazim removal from wastewater runoff. A two-stage pilot CW was built and tested for its efficacy of carbendazim removal under saturated conditions and varying hydraulic loading rates. The influent was pumped into the first vertical-flow mesocosm. The drained water was then pumped into the second mesocosm. The collected effluent was tested for carbendazim removal. Carbendazim removal efficiencies up to 91.80% (with a hydraulic loading rate of 100 Ld^{-1} and influent carbendazim concentration of 10 μg L^{-1}) were observed. Statistical analysis indicated that the removal of carbendazim was not correlated with the initial carbendazim concentration but was negatively correlated with the hydraulic loading rate used. Two pesticide removal mechanisms were briefly probed to determine their participation in carbendazim removal. Substrate sorption accounted for 18% of total carbendazim removal; furthermore, plant uptake also played an active role.

Key words: carbendazim, constructed wetlands, floriculture, vertical flow constructed wetland

HIGHLIGHTS

- Novel two-stage pilot-scale constructed wetland was built to reflect industry standard configuration of constructed wetlands.
- Tests under saturated conditions showed up to 91.80% carbendazim removal from the influent water.
- Carbendazim removal efficiency was found to be positively correlated with a higher hydraulic retention time but to be independent of the influent carbendazim concentration.

GRAPHICAL ABSTRACT

Carbendazim Remediation using Constructed Wetlands

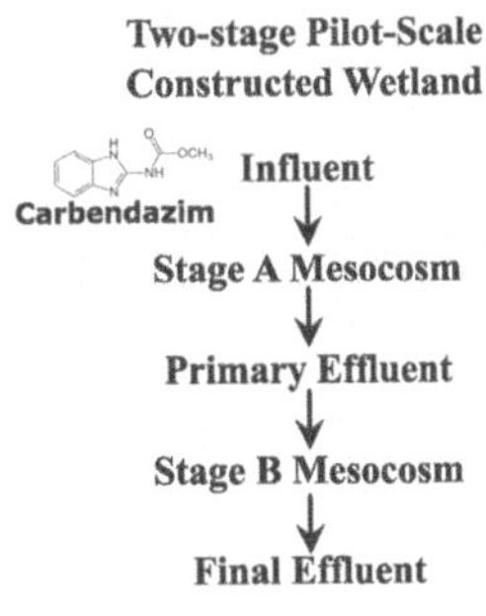

INTRODUCTION

In fast-growing economies, the sufficient and permanent supply of resources is a great challenge for governmental entities, and treatment of polluted resources and protection of natural water bodies present an even greater issue. In the field of wastewater treatment, polluted water requires additional energy for treatment. Planning of the required volumes and quality of wastewater is complicated (especially in industrial areas) and subject to changes in short periods. In contrast to the supply of water to the economy, which is essential for its operation, the treatment of polluted resources has no direct benefit for the consumers and only adds additional cost. Thus, treatment of polluted resources reduces the profitability of production. Lower environmental standards give economies a monetary advantage on the global market. The negative effects of such a situation can take a long time to manifest as severe problems for humans and the environment.

The floriculture industry is booming in Ethiopia and was responsible for exports worth USD 188.5 million during 2019. Flowers are luxurious products with high social and economic value, and Ethiopia is ranked second in Africa and sixth amongst the top exporters of flowers worldwide (EHPEA 2019). Increasing demand drives a growing dependence on pesticides in large-scale flower farms to increase the rate of production. The flower farms are generally situated next to water bodies (EHPEA 2018), and the residual pesticides find their way into these water bodies through runoffs from these farms. In recent years, several initiatives including constructed wetlands (CWs) have been brought forward to combat this situation. Nevertheless, improved studies on the effective treatment of pesticide effluents are needed for larger-scale design and implementation.

One of the commonly used pesticides in Ethiopian flower farms is carbendazim. It is found in two registered and commercially available products, one of which is specifically used in flower farms (Plant Regulatory Directorate 2018). Residual carbendazim is shown to have a half-life of anywhere between 3 days to 12 months based on the local environment. This is ample time for it to enter the ecosystem and accumulate. Eventually, it will cross the threshold concentrations and start affecting the surrounding flora and fauna. Moreover, carbendazim has been classified as a possible human carcinogen and is a known endocrine-disrupting chemical. It is also known to manifest embryotoxicity, infertility, and developmental toxicity in mammals (Singh *et al.* 2016). For these reasons, the use of carbendazim is already banned in the European Union (Commission Implementing Regulation (EU) No 542/2011, 2011). However, this compound has been recently detected in surface water (at a sub-micro gL^{-1} level) in Spain (Calvo *et al.* 2021).

Removal of pollutants such as carbendazim from water sources is a great challenge for developing countries. Construction and operation of conventional water treatment plants designed specifically for treating these pollutants is a less desirable solution as they are costly, sometimes require high energy consumption, and need imports of specialised materials. CWs mimic the functions of natural wetlands and have been engineered to treat raw sewage and agricultural and industrial effluents (Vymazal 2014; Vymazal & Březinová 2015). They also do not have high energy requirements and have low operational costs. Awareness and implementation of CWs has been increasing worldwide over the years and they have been introduced in many large projects such as the World Expo 2020 in Dubai, United Arab Emirates.

The use of CWs for the treatment of pesticide effluents has been studied extensively over the years. While their use in the context of pesticide removal was first studied in the 1970s, it is only in recent years that CWs have gained widespread use for pesticide removal from agricultural runoffs (Vymazal & Březinová 2015). A few pesticides which have been effectively treated using CWs include chlorpyrifos, chlorothalonil, imazalil, and tebuconazole (Sherrard *et al.* 2004; Lv *et al.* 2016; Tang *et al.* 2019). However, the treatment of carbendazim using CWs has not been specifically studied in depth compared to the examples stated previously (Kang *et al.* 2020; McCalla *et al.* 2022). Thus, there is a need for further investigation to elucidate whether CWs can be a viable option for effective treatment of agriculture and floriculture runoffs which contain carbendazim.

Two-stage vertical-flow CWs have been used for sewage water treatment. The German Association for Water, Wastewater and Waste (DWA) has also described design guidelines for the construction and operation of two-stage vertical-flow CWs (DWA-A 262E 2017). However, there have not been many studies related to the use of the same for pesticide removal from agriculture and floriculture runoffs. It is an interesting prospect to study the effectiveness of a two-stage vertical-flow CW for pesticide removal. There is also a need for developing engineered and scalable CW solutions which are specifically suitable for use in Ethiopian flower farms.

This study aims at studying the removal efficiencies of carbendazim at different inflow concentrations using a pilot-scale vertical flow CW and the effect of varying hydraulic loading rate (HLR) and inflow carbendazim concentration on the removal efficiencies of carbendazim. This study also briefly analyses the mechanisms involved in the removal of carbendazim from influent water.

MATERIALS AND METHODS

Materials

Phragmites australis (common reed) was selected as the wetland plant species to be used in this experiment. Saplings were sourced from the nursery operated and maintained by Reed Bed Waste Water Treatment Company in Sharjah, United Arab Emirates. Each mesocosm was constructed using an Immediate Bulk Container ($117 \times 95 \times 100$ cm). Double-washed filter material of three different particle sizes was used in the construction of the mecocosm – double-washed black fine gravel (grain size 1–5 mm), smaller gravel (grain size 10–15 mm), and larger gravel (grain size 20–25 mm).

Experimental setup

The CW mesocosms were constructed to mimic large-scale CWs built following the example of French vertical-flow wetlands (Dotro *et al.* 2021) and guidelines published by the DWA (DWA-A 262E 2017). However, contrary to the textbook and the guidelines, only one filter unit was used for the first stage (Stage A), instead of the recommended minimum of three filter units. Another point to note was that the filter material used in the Stage A mesocosms was the same as the one used in the second stage (Stage B) mesocosms. These changes were made based on the assumption that the raw wastewater is free from solids, and therefore does not require filtration with fine gravel and a resting period in stage A.

The construction site was located at the RAK Research and Innovation Center, Ras Al Khaimah, U.A.E. The experiments were run from October 3, 2021 to December 30, 2021, and the average air temperature was in the range of 23–49 °C during this period. There were mostly sunny days during this period due to which precipitation was assumed to have a negligible impact on the operation. Stage A mesocosms were filled with a 20 cm layer of larger gravel at the bottom, a 10 cm layer of smaller gravel above it, and a 40 cm layer of fine gravel on top. Stage B mesocosms were filled with a 20 cm layer of larger gravel at the bottom, a 10 cm layer of smaller gravel above it, and a 60 cm layer of fine gravel on top. The outlet was placed 5 cm below the surface level to avoid photodegradation of carbendazim as well as to keep the mesocosm constantly saturated with influent water. Three treatment groups were set up: one for control (no carbendazim) and two for testing two different influent concentrations (10 and 100 µg L^{-1}) of carbendazim. For testing under unsaturated conditions, the level of the outlet was adjusted to the bottom of the mesocosm and testing was done using two treatment groups: one for control and one for 100 µg L^{-1} influent concentration of carbendazim.

For each group, similarly sized plants were selected randomly and planted following a planting grid in the mesocosms. The influent was pumped from the influent tank to the Stage A tank. The effluent was then collected into an underground tank to be further pumped to the Stage B tank. The final effluent was then collected into the last collection tank. Figure 1 shows a detailed schematic diagram of the pilot-scale CW system constructed for this study. The HLR of the influent, located in chamber A, was timer-controlled. The HLR of the Stage A effluent (located in chamber B) was adjusted to match the influent HLR with the help of float switches. Before the start of experimentation, treated sewage effluent water was introduced to the system to facilitate plant growth and to introduce a microbial population.

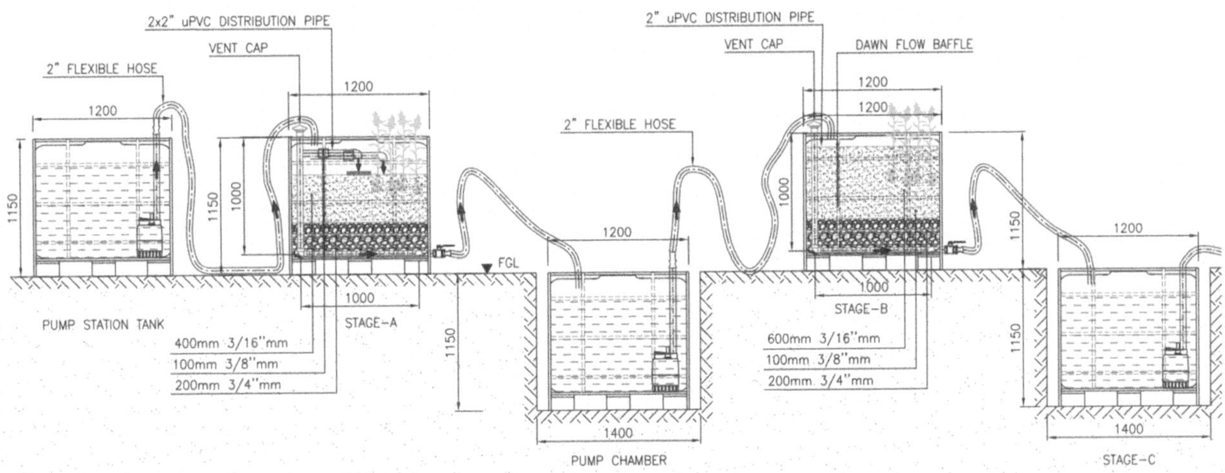

Figure 1 | Schematic diagram of pilot-scale Constructed Wetland.

The influent was prepared in batches of 500 L. For preparation of the influent, concentrated solutions of carbendazim were first prepared in the lab by dissolving carbendazim in 200 mL of methanol. Absolute methanol was used because carbendazim has very low solubility in water. The addition of methanol also served as an additional carbon source to be added to the influent water. The influent tanks of each stream were filled with 500 L of water, to which the prepared concentrates were added. In this study, we used an HLR of 400, 200, and 100 L d^{-1} to study the effect of carbendazim concentrations as well as HLR on the removal efficiency of carbendazim. The average flow capacity of the pumps used was calculated to be 3.5 L s^{-1}. Subsequently, the activation time of influent pumps was calculated to be 19 s (HLR of 400 L d^{-1}), 10 s (HLR of 200 L d^{-1}) and 5 s (HLR of 100 L d^{-1}) every 4 hours.

Sample collection was done through taps placed at the outlets of each Stage B tank. Samples of influents were collected directly from the influent tank; a volume of 1 L was collected for each sample of influent and effluent. The volume of total effluent collected at the end of Stage B was recorded to account for water loss due to evapotranspiration. Sampling of influent was carried out at the beginning of the experiment while sampling of effluent was carried out at the end of the experiment. Samples were stored in a 4 °C refrigerator before being processed to calculate the carbendazim removal efficiency of the system. Other parameters measured immediately after sample collection included pH, temperature, dissolved oxygen (DO), and electrical conductivity (EC). Samples were also collected to measure the biochemical oxygen demand (BOD), chemical oxygen demand (COD), Total Organic Carbon (TOC), nitrates, and phosphates for the control and 100 µg L^{-1} groups (under saturated conditions). Plant samples were collected by harvesting the plant stem from 10 cm above the gravel level of the mesocosm. This way, the carbendazim content analysis for the plant sample would represent the amount of carbendazim absorbed by the plant.

Analytical methods

Carbendazim concentration analysis

Sample preparation was carried out using the liquid-liquid extraction method to obtain an extracted sample concentrate of 1 mL. The pH of the sample was first adjusted to 1–2, followed by separation of the aqueous and organic phases. Organic content of the sample was then extracted using dichloromethane, followed by filtration using sodium sulphate. The leftover solution was centrifuged and then concentrated to 1 mL using nitrogen gas. It was then processed using gas chromatography-mass spectrometry (GC-MS) along with the standard solutions to calculate the concentration of carbendazim in the sample with an accuracy of 0.01 µg L^{-1} (Shin *et al.* 2001).

For extraction of carbendazim from substrate sample (gravel), 10 g of the sample was first placed in a polycarbonate tube. Subsequently, 10 g of anhydrous sodium sulphate and 25 mL of methanol were added to this tube, and the tube was subjected to shaking and sonification for 15 min. This resulted in a separated organic layer, which was evaporated under a nitrogen stream. Then, 50 µL of derivatising agent (iodomethane) was added to the collected residue and heated at 90 °C for 60 min. Subsequently, 100 µL of ethyl acetate was added to the residue to make the sample ready for GC-MS analysis (Shin *et al.* 2001).

Method 8081B described by the US EPA was used (US EPA 2015) for extraction of carbendazim from plant sample. The GC-MS instrument used for all the quantification procedures is Shizmadu GCMS-QP2010 SE (manufactured by Shimadzu (Suzhou) Instruments Manufacturing Co., Ltd, Jiangsu, China) and the standard dilutions used to generate the calibration curve were 0.01, 0.1, 1, 10, 100, and 1,000 µg L^{-1}. Spiking of processed samples with known carbendazim concentrations was also performed and the results of both spiked and non-spiked samples were compared to assure the quality of the results obtained.

Removal mechanisms for carbendazim

Two of the many possible carbendazim removal mechanisms (Imfeld *et al.* 2009) were considered in this study to check whether or not these mechanisms play an active role in carbendazim removal from influent water. The first mechanism was substrate sorption and the second was plant uptake.

Double washed fine gravel used in the construction of the CWs was taken and placed in a clean plastic container to evaluate the role of substrate sorption in carbendazim removal. This container was then filled with carbendazim water with a carbendazim concentration of 2 g L^{-1} (high carbendazim concentration was used to make sure that there would be detectable amounts of carbendazim adsorbed by the substrate). This was then left undisturbed and kept away from sunlight for 4 days. The water was then collected and analysed for the concentration of carbendazim present in the water. The gravel was also processed and analysed to determine the sorption of carbendazim.

Plant samples were taken from the mesocosm, which was dosed with 100 µg L^{-1} concentration of carbendazim, and analysed for the presence of carbendazim in them to investigate its removal through plant uptake.

Methods of calculation

Each mesocosm was saturated with water before cutting off the water supply prior to running the experiments. Subsequently, the amount of effluent collected from each mesocosm was measured, which was used to calculate the hydraulic retention time (HRT) of the mesocosm. The HRT was calculated as follows:

$$T_r = V_{out}/Q \tag{1}$$

where T_r is the HRT, V_{out} the volume of effluent, and Q is HLR in L d^{-1}.

Evapotranspiration was calculated as follows:

$$\triangle V = (Qt - V_{out})/Qt \tag{2}$$

where $\triangle V$ is the water loss by evapotranspiration, Q is the influent flow rate in L d^{-1}, t is the time in days, and V_{out} is the total effluent volume recorded.

This value was used to correct the calculation of removal efficiencies of the carbendazim. The removal efficiencies were calculated as follows:

$$\text{Removal efficiency } (\%)(C_{in} - (1 - \triangle V) \times C_{out})/C_{in} \times 100 \tag{3}$$

where C_{in} and C_{out} are influent and effluent concentrations, respectively (in µg L^{-1}).

Based on previous pesticide-related studies, carbendazim removal is expected to follow a first-order kinetic model (Lyu *et al.* 2018). The fitted values for the model are derived using the formula:

$$k = ln\left(\frac{C_{in}}{C_{out}}\right)/T_r \tag{4}$$

where k is the first-order rate constant (in d^{-1}), C_{in} and C_{out} are influent and effluent concentrations, respectively (in µg L^{-1}), and T_r is the HRT.

Data analysis

Statistical analysis as well as two-way analysis of variance (ANOVA) were carried out using JASP software (JASP Team 2022). Two-way ANOVA was used to identify the effect of influencing factors (HLR and influent carbendazim concentration) on the first-order rate constant, k, for carbendazim removal from the influent water. It was carried out at a significance level of 0.05 ($p<0.05$). Data plots were generated with the help of Veusz software (Sanders *et al.* 2021).

RESULTS

Carbendazim removal

Carbendazim levels were measured and removal efficiencies were calculated using Equation (3). As expected, the carbendazim levels observed in the control were below the detection limit. The removal efficiency of carbendazim as well as the corresponding first-order rate constant, k, were calculated for three different hydraulic loading rates: 100, 200, and 400 L d^{-1}. The results are depicted in Table S1 in the Supplementary data. Carbendazim removal efficiencies were observed in the range of 68.87–91.80%, with higher removal efficiencies at HLR of 100 L d^{-1} and lower removal efficiencies at HLR of 400 L d^{-1}.

A graphical representation of this data is shown in Figure 2. The removal efficiency showed a decreasing trend as the HLR increased. This could be attributed to a higher HLR which implies a lower retention time in the system. Further, this means that contact time has a notable influence on the effective removal of carbendazim in the CW system employed in this study. To try to compare these findings with unsaturated conditions, the experiment was repeated at an HLR of 200 L d^{-1} and with an influent carbendazim concentration of 100 µg L^{-1}. The results showed an average carbendazim removal efficiency of 16.14% under unsaturated conditions with an average k-value of 1.7860 d^{-1}. The HRT observed during this experiment was 0.0833 days (2 h).

A comparison of the first-order rate constants, k, observed for each carbendazim concentration at different HLRs under saturated conditions indicated that the observed k-values seemed to be quite similar to each other. This could imply that the first-order rate constant, k, is independent of the influent concentration of carbendazim used.

To further confirm this observation statistically, we carried out a two-way ANOVA to observe the effect of HLR and carbendazim concentration on the first-order rate constant, k, for the removal of carbendazim from influent water under saturated conditions. A significance level of 95% was used and the result can be seen in Table 1 The p-value for the effect of HLR on k was below 0.05, which means it is a significant result. The p-value for the effect of carbendazim concentration was above 0.05. This means that the difference in k-values between the two influent concentrations, at the same HLR, was not significant. The p-value for the interactive effect of HLR and carbendazim concentration on k was above 0.05. This means that it is not a significant result, implying that k is independent of any interaction between these variables.

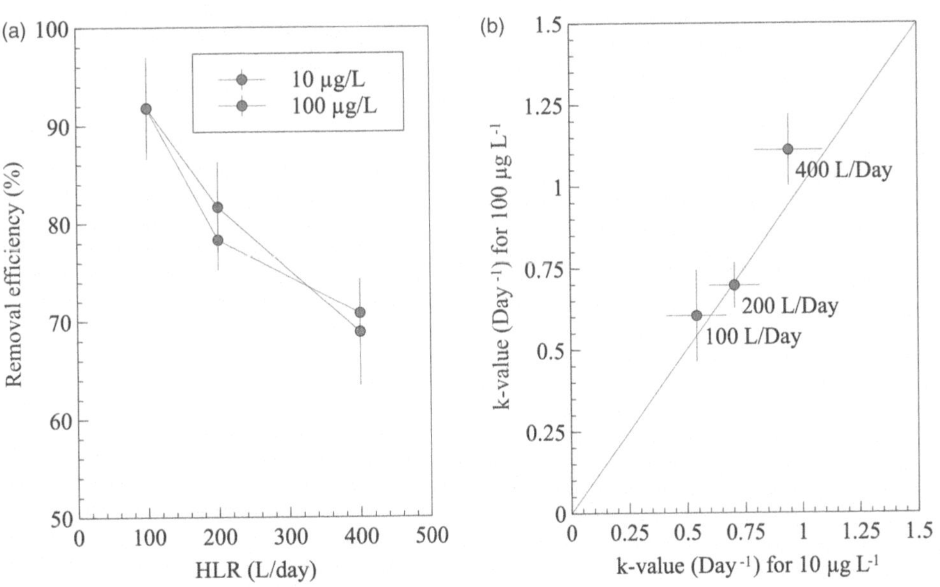

Figure 2 | (a) Comparison of carbendazim removal efficiencies observed at different HLRs for two different influent concentrations, under saturated conditions. (b) Comparison of first-order rate constants, k, observed at different HLRs for the two different influent carbendazim concentrations. The error bars used in both the plots represent the standard deviations of the data observation.

Table 1 | Two-way ANOVA analysis to verify the effect of influent carbendazim concentration, C_{in}, and HLR on the first-order rate constant, k, for the removal of carbendazim from influent water

	Cases	Sum of Squares	df	Mean Square	F	P
a.	**Between Subjects Effects**					
	HLR	0.658	2	0.329	16.709	0.004
	Residuals	0.118	6	0.020		
b.	**Within Subjects Effects**					
	C_{in}	0.025	1	0.025	2.530	0.163
	C_{in} * *HLR*	0.023	2	0.012	1.209	0.362
	Residuals	0.058	6	0.010		

Note. Type III Sum of Squares.

Part (a) illustrates the results for the effect of HLR on k. Part (b) illustrates the results for the effect of C_{in} and C_{in} * HLR on k.

Water parameters

Other parameters which were monitored to observe variations in the experimental conditions as well as to evaluate the effect of treatment by CWs on the water quality included pH, DO, EC, and water temperature (Table 2). Average pH of the influent was observed to be in the range of 7.58–7.71, while that of the effluent was in the range of 9.47–9.57 across all three experimental groups. The percentage saturation of dissolved oxygen was observed to be $\approx 21\%$ for the influent and $\approx 32\%$ for the effluent. The electrical conductivity for the influent was found to be ≈ 440 uS cm^{-1} and that of the effluent was found to be ≈ 825 uS cm^{-1}. The average temperatures of all influent and effluent samples were found to be ≈ 37 °C.

Another set of parameters which were measured for the control and 100 µg L^{-1} samples were BOD, COD, TOC, nitrates, and phosphates (Table 3). A higher level of COD was observed for the influent of the 100 µg L^{-1} carbendazim experimental group. This could be attributed to the fact that methanol was used as part of influent preparation for this experimental group while it was not used in the control group.

Removal mechanics

The role of substrate sorption as a means of removal of carbendazim was studied. The substrate sample was collected post-exposure to carbendazim (as described in the methods section) and was analysed for the carbendazim content present in it. The results obtained showed that substrate sorption was responsible for the removal of 18% of the total carbendazim content present in the prepared influent water (Table 4). This implies that substrate sorption takes an active part in removal of carbendazim from the influent water.

Plant samples collected from the mesocosms were analysed for their carbendazim content to test the possibility of carbendazim removal via plant uptake. Carbendazim content in the plant sample was observed to be 5.42 mg kg^{-1}. This value is representative of the total amount of carbendazim absorbed by the plant over the entire duration of the experimentation and does not account for the amount that could have been possibly degraded by the plant. Although this does not indicate the relative contribution of plant uptake in carbendazim removal, the presence of carbendazim in plant sample is proof

Table 2 | Data from water parameters tested immediately after sampling from the mesocosms

		pH		DO (%)		EC (uS cm^{-1})		Temperature (°C)	
Control	*Influent*	7.58	±0.175	20.8	±2.201	440.2	±25.328	36.6	±1.075
	Effluent	9.54	±0.184	32.9	±2.807	821.2	±13.790	36.6	±1.265
10 µg L^{-1}	*Influent*	7.74	±0.104	21.6	±2.171	444.3	±30.930	36.4	±1.265
	Effluent	9.47	±0.223	31.6	±2.503	824.4	±18.057	37.5	±0.972
100 µg L^{-1}	*Influent*	7.71	±0.078	21.6	±2.633	444.2	±24.303	37.1	±1.912
	Effluent	9.57	±0.162	34.0	±3.018	830.5	±18.435	37.1	±1.197

The parameters tested were pH, dissolved oxygen (DO, measured as percentage saturation), electrical conductivity (EC, measured as uS cm^{-1}) and temperature (measured in °C).

Table 3 | Data for BOD, COD, TOC, nitrates, and phosphates parameters observed for the control sample and sample from 100 μg L^{-1} dosed experimental groups

		BOD (mgL^{-1})	COD (mgL^{-1})	TOC (mgL^{-1})	Nitrate (mgL^{-1})	Phosphate (mgL^{-1})
Control	*Influent*	<12	11	0.68	0.05	0.36
	Effluent	<12	<10	0.28	0.06	<0.02
100 μg L^{-1}	*Influent*	<12	1302	260	0.04	0.07
	Effluent	<12	14	4.72	0.05	<0.02

Table 4 | Observations and calculations made to calculate the contribution of substrate sorption towards removal of carbendazim from prepared influent water (concentration of 2 mg L^{-1})

Total amount of carbendazim added to container	10 mg
Total amount of soil present in the container	2 kg
Carbendazim concentration observed in the substrate at the end of the experiment	0.9 mg kg^{-1}
Amount of carbendazim removed by substrate sorption	1.8 mg

that carbendazim is absorbed by plants and that plant uptake plays an active role in the carbendazim removal from the influent water.

DISCUSSION

It is noteworthy that the design of the pilot CW used in this study is built with the example of French wetlands and DWA guidelines (DWA-A 262E 2017; Dotro *et al.* 2021) in mind, and is meant to mimic actual field conditions as much as possible. There have been previous studies that make use of pilot-scale CWs for the removal of pesticides (Lyu *et al.* 2018), but not many use a two-stage vertical-flow design that mimics how large scale CWs are built commercially. This design also allows for easy modifications to the system for future studies with varied experimental requirements, for example, the use of effluent recirculation and oxygenation among others. This allows the study results to be easily scalable for industrial purposes.

The use of a saturated system has shown great promise, with carbendazim removal efficiencies of up to 91.80%. The use of a completely unsaturated system showed a much lesser efficiency of 16.14% at an HLR of 200 L d^{-1}, while the corresponding removal efficiency in the saturated system was 78.9%. This could be attributed to its drastically low retention time as compared to a saturated system. While the HRT for a saturated system with an HLR of 200 L d^{-1} was 2.11 days, the corresponding HRT for an unsaturated system was 0.0833 days (2 h). Taking this into consideration, we could achieve much higher removal efficiencies if we could increase the retention time of the carbendazim water while maintaining an unsaturated condition. One way to address this would be recirculating the effluent water resulting in greater retention of the water in the system. If we were to recirculate the effluent water four times after initial loading, we could potentially achieve a matching removal efficiency as the saturated system. The corresponding retention time of the water would be 0.417 days; thus, the retention would be five times faster than the saturated system. This warrants further investigation as it could allow for use of higher HLR to increase the efficiency of the CWs as a whole.

Additionally, there was an increase in the pH of the effluent water as compared to that of the influent water. On average, the pH of the effluent water was above 9 across the three streams used in this study. This makes it less suitable for applying the treated water for irrigation purposes or simply allowing it to enter nearby water bodies (Jeong *et al.* 2016). During our experimental observations, pH values for effluent water under unsaturated conditions were significantly less than 9. This implies that the increase in observed pH was positively correlated to an increase in the saturation of the system. Thus, further research is required for optimization of the saturation level to achieve maximum pesticide removal without compromising effluent water quality. The use of a partially saturated system would also theoretically increase dissolved oxygen levels.

This could also lead to a possible increase in the removal efficiency (Vink & van der Zee 1997), but it needs further investigation.

While substrate adsorption was determined to be responsible for 18% of the carbendazim removal, the degree of involvement of plant uptake needs further in-depth study. Although removal mechanisms were evaluated in this study, they were studied separately. All the possible removal mechanisms for carbendazim are highly complex in nature and are interdependent upon each other as well. For example, over time there tends to be a build-up of biofilm in the gravel used in the CWs. This may decrease the amount of substrate sorption but the role of the microbial population in the degradation of carbendazim could enhance (McBride & Tanner 1999; Chyan *et al.* 2013). This is why there is a need to further study the correlation of the various possible removal mechanisms as well. This work would also help illustrate the mass balance of the carbendazim introduced to the constructed wetland.

CONCLUSION

The results from this study indicate that a two-stage vertical-flow CW system is an effective system for the removal of carbendazim from wastewater runoffs from floriculture farms. Moreover, there is a need for further optimisation of the CW system to achieve maximum performance without compromising the effluent water quality for discharge into water bodies.

Removal efficiencies of up to 91.80% were observed under saturated conditions. Further analysis using two-way ANOVA showed that the first-order rate constant, k, for removal of carbendazim was affected by the HLR used, but was unaffected by the influent carbendazim concentration or any interactive effect between HLR and influent carbendazim concentration. Comparison of the saturated vs. unsaturated system indicated great potential for testing the unsaturated system with recirculation of effluent to effectively increase the efficiency of carbendazim removal from the influent water.

Adsorption to the substrate and plant uptake played a significant role in the removal of carbendazim from the wastewater. However, further studies are warranted to identify the relative contributions and interdependence of these removal mechanisms and others such as biodegradation pathways.

ACKNOWLEDGEMENTS

This paper and research would not have been possible without the exceptional support of our research supervisors, Dr Feleke Zewge and Dr Yoshihiko Inagaki. Their extensive knowledge and experience were a great inspiration to all of us involved in this research. We would like to thank Dr Uday Kumar from American University of Ras Al Khaimah, United Arab Emirates. Our research would not have been possible without his support in arranging a site for the construction of pilot CWs as well as providing access to laboratory equipment needed for the research. We would like to thank Mr Wolfram Sievert from Sievert Consult, Lemgo, Germany for his support in designing the pilot CWs. We would also like to thank Mr Akshay Deshpande from Reed Bed Waste Water Treatment, Dubai, United Arab Emirates for his on-site assistance during the experimentation. We are also grateful to the anonymous peer reviewers for their insightful comments and suggestions. Their contribution has helped improve the coherence and quality of this paper.

DATA AVAILABILITY STATEMENT

All relevant data are included in the paper or its Supplementary Information.

CONFLICT OF INTEREST

The authors declare there is no conflict.

REFERENCES

Calvo, S., Romo, S., Soria, J. & Picó, Y. 2021 Pesticide contamination in water and sediment of the aquatic systems of the Natural Park of the Albufera of Valencia (Spain) during the rice cultivation period. *Science of the Total Environment* **774**, 145009.

Chyan, J.-M., Senoro, D.-B., Lin, C.-J., Chen, P.-J. & Chen, I.-M. 2013 A novel biofilm carrier for pollutant removal in a constructed wetland based on waste rubber tire chips. *International Biodeterioration & Biodegradation* **85**, 638–645.

Commission Implementing Regulation (EU) No 542/2011 2011 *Official Journal of the European Union*. Available from: https://eur-lex.europa.eu/eli/reg_impl/2011/542#ntc2-L_2011153EN.01018901-E0002 (accessed 23 November 2021).

Dotro, G., Langergraber, G., Molle, P., Nivala, J., Puigagut, J., Stein, O. & von Sperling, M. 2021 *Treatment Wetlands*. IWA Publishing, London, UK.

DWA-A 262E 2017 *Principles for Dimensioning, Construction and Operation of Wastewater Treatment Plants with Planted and Unplanted Filters for Treatment of Domestic and Municipal Wastewater*. DWA Set of Rules, German Association for Water, Wastewater and Waste (DWA), Hennef, Germany.

EHPEA 2018 *EHPEA | Farm Locations*. Available from: https://ehpea.org/farm-location/ (accessed 9 March 2022).

EHPEA 2019 *Ethiopia Earned 230 Million USD From Horticulture Export – EHPEA*. EHPEA. Available from: https://ehpea.org/ethiopia-earned-230-million-usd-from-horticulture-export/ (accessed 16 March 2022).

Imfeld, G., Braeckevelt, M., Kuschk, P. & Richnow, H. H. 2009 Monitoring and assessing processes of organic chemicals removal in constructed wetlands. *Chemosphere* **74** (3), 349–362.

JASP Team 2022 *JASP (Version 0.16.1) [Computer Software]*. Available from: https://jasp-stats.org/ (accessed 18 March 2022).

Jeong, H., Kim, H. & Jang, T. 2016 Irrigation water quality standards for indirect wastewater reuse in agriculture: a contribution toward sustainable wastewater reuse in South Korea. *Water* **8** (4), 169.

Kang, D., Doudrick, K., Park, N., Choi, Y., Kim, K. & Jeon, J. 2020 Identification of transformation products to characterize the ability of a natural wetland to degrade synthetic organic pollutants. *Water Research* **187**, 116425.

Lv, T., Zhang, Y., Zhang, L., Carvalho, P. N., Arias, C. A. & Brix, H. 2016 Removal of the pesticides imazalil and tebuconazole in saturated constructed wetland mesocosms. *Water Research* **91**, 126–136.

Lyu, T., Zhang, L., Xu, X., Arias, C. A., Brix, H. & Carvalho, P. N. 2018 Removal of the pesticide tebuconazole in constructed wetlands: design comparison, influencing factors and modelling. *Environmental Pollution* **233**, 71–80.

McBride, G. B. & Tanner, C. C. 1999 Modelling biofilm nitrogen transformations in constructed wetland mesocosms with fluctuating water levels. *Ecological Engineering* **14** (1–2), 93–106.

McCalla, L. B., Phillips, B. M., Anderson, B. S., Voorhees, J. P., Siegler, K., Faulkenberry, K. R., Goodman, M. C., Deng, X. & Tjeerdema, R. S. 2022 Effectiveness of a constructed wetland with carbon filtration in reducing pesticides associated with agricultural runoff. *Archives of Environmental Contamination and Toxicology*. Available from: https://link.springer.com/10.1007/s00244-021-00909-0.

Plant Regulatory Directorate. 2018 *List of Registered Pesticides in Ethiopia*.

Sanders, J., Bell, G., Graham, J., Harris, B., Hughes, D., Mussi, V. & Stuhl, B. K. 2021 *Veusz*. Available from: https://veusz.github.io/ (accessed 1 June 2022).

Sherrard, R. M., Bearr, J. S., Murray-Gulde, C. L., Rodgers, J. H. & Shah, Y. T. 2004 Feasibility of constructed wetlands for removing chlorothalonil and chlorpyrifos from aqueous mixtures. *Environmental Pollution* **127** (3), 385–394.

Shin, H. S., Hong, J. E., Pyo, H., Park, S.-J. & Lee, W. 2001 Sensitive determination of benomyl in environmental samples by GC/MSD. *Analytical Sciences/Supplements* **17asia**, a49–a52.

Singh, S., Singh, N., Kumar, V., Datta, S., Wani, A. B., Singh, D., Singh, K. & Singh, J. 2016 Toxicity, monitoring and biodegradation of the fungicide carbendazim. *Environmental Chemistry Letters* **14** (3), 317–329.

Tang, X. Y., Yang, Y., McBride, M. B., Tao, R., Dai, Y. N. & Zhang, X. M. 2019 Removal of chlorpyrifos in recirculating vertical flow constructed wetlands with five wetland plant species. *Chemosphere* **216**, 195–202.

US EPA 2015 *Method 8081B*. Available from: https://www.epa.gov/sites/default/files/2015-12/documents/8081b.pdf (accessed 17 March 2022).

Vink, J. P. M. & van der Zee, S. E. A. T. M. 1997 Effect of oxygen status on pesticide transformation and sorption in undisturbed soil and lake sediment. *Environmental Toxicology and Chemistry* **16** (4), 608–616.

Vymazal, J. 2014 Constructed wetlands for treatment of industrial wastewaters: a review. *Ecological Engineering* **73**, 724–751.

Vymazal, J. & Březinová, T. 2015 The use of constructed wetlands for removal of pesticides from agricultural runoff and drainage: a review. *Environment International* **75**, 11–20.

First received 31 March 2022; accepted in revised form 21 June 2022. Available online 28 June 2022

doi: 10.2166/wst.2022.191

Assessing the impact of micropollutant mitigation measures using vertical flow constructed wetlands for municipal wastewater catchments in the greater region: a reference case for rural areas

Silvia Venditti [a],*, Anne Kiesch [b], Hana Brunhoferova [a], Markus Schlienz[a], Henning Knerr[c], Ulrich Dittmer[c] and Joachim Hansen [a]

[a] Chair of Urban Water Management, University of Luxembourg, 6, rue Coudenhove-Kalergi, L-1359, Luxembourg, Luxembourg
[b] Previously Chair of Urban Water Management, University of Luxembourg, 6, rue Coudenhove-Kalergi, L-1359, Luxembourg, Luxembourg, now TR-Engineering, 86-88, Rue de l' Egalité, L-1456, Luxembourg, Luxembourg
[c] Department of Urban Water Management, University of Kaiserslautern, Paul-Ehrlich-Straße 14, D- 67663, Kaiserslautern, Germany
*Corresponding author. E-mail: silvia.venditti@uni.lu

SV, 0000-0003-1775-8371; AK, 0000-0003-2734-5693; HB, 0000-0001-6582-9803; JH, 0000-0003-4476-4849

ABSTRACT

The present research aims at giving an approach to the issue of surface water contamination due to micropollutants in rural areas. The catchment of the Sûre river was selected as a reference case for the Greater Region, characterized mainly by settlements with low population density, small water bodies and small- to medium-sized wastewater treatment plants (WWTPs). For these WWTPs, conventional technical solutions for micropollutant elimination are not suitable; therefore, an adapted mitigation strategy is needed to prevent the impact of micropollutants, especially during the dry season. As a suitable alternative to more intensive technologies, Constructed Wetlands (CW) in Vertical Flow (VF) configuration have been successfully tested over a 1-year period and the elimination rate of 27 micropollutants was quantified. Emission reduction by VF was then considered in a static mass balance model that calculates the longitudinal concentration profile for the entire river catchment. The EmiSûre approach, which focuses on river quality (concentrations of pollutants) instead of emitted loads, effectively allowed simulation of adopted measures *a priori* and resulted in efficient support for decision-makers with WWTP upgrade scenarios.

Key words: constructed wetlands, emission measures, EmiSûre model, micropollutants

HIGHLIGHTS

- A novel approach is applied to assess mitigation measures for micropollutants elimination in rural areas.
- Constructed wetlands are promising for micropollutants elimination.
- Emission reductions by Constructed Wetlands are considered in a static mass balance model.
- The selection of the WWTPs resulted in a significant load reduction.
- The EmiSûre model successfully serves as a valuable tool for decision-makers.

INTRODUCTION

Although in the last twenty years the need to minimize micropollutant emission has found a consensus in the upgrade of conventional Wastewater Treatment Plants (WWTPs) with advanced technologies (Eggen *et al.* 2014; Falås *et al.* 2016), a proper solution to the way in which small and medium-sized WWTPs in rural settlements should be upgraded is still needed.

Mitigation strategies have been focused mainly on large WWTPs where the highest ratio of mass reduction to cost (cost-effectiveness) can be applied with end-of-pipe solutions. For those existing WWTPs, the implementation of intensive technologies has been demonstrated to lead to high long-term investment with consequent additional energy consumption (UBA 2009).

Micropollutants released from small WWTPs located in rural areas can also lead to critical concentrations in the receiving waters mainly due to the limited dilution factor and eventually to the more vulnerable hydrology of the rivers. Surface waters in rural areas often present concentrations of pollutants exceeding the Environmental Quality Standards (EQSs) values, which makes it necessary to introduce measures for achieving and maintaining *good water status* as the main principle of the Water Frame Directive (EC 2013).

In this paper, we propose an approach to such cases and particularly urgent for the Greater Region (German federal states Rhineland-Palatinate and Saarland, the Grand Duchy of Luxembourg, regions Wallonia and Lorraine from Belgium and France, respectively) characterized mainly by rural areas with low population density and small water bodies with limited dilution, especially during summer. Among the possible rivers, the Sûre has been selected as a representative case, being the border between Luxembourg and Germany and connecting a total of 286 WWTPs in its catchment.

Because intensive technologies are not affordable to be applied to small WWTPs, nature-based solutions such as Constructed Wetlands (CWs) are here considered as a potential alternative for micropollutants removal.

The complexity of the multiple mechanisms (i.e. sorption, photodegradation, phytodegradation, and bioremediation) acting together in the removal of pollutants and the fact that the efficiency can be enhanced via the targeted influence of some of them (i.e. selection of substrate to influence sorption) has been an object of previous studies (Li *et al.* 2014; Ilyas & van Hullebusch 2019; Ilyas *et al.* 2020). However, the lack of data available for their application as post-treatment did not help to trust their viability.

The main objective of this study is thus to assess the impact of micropollutant removal measures in the Sûre catchment when post-treatment steps are applied and to offer a model approach for catchments of similar characteristics.

A pilot case study with CW in Vertical Flow (VF) configuration was conducted to define the occurrence of micropollutants for a WWTP that collects industrial and domestic streams from both Germany and Luxembourg and to assess its removal efficiencies as those achieved with the implementation of the post-treatment step.

Emission reduction by VF was thus considered for WWTP <10,000 PE in a mass balance model (named EmiSûre model) that calculates longitudinal concentrations profile for the entire river catchment by superimposing the point emissions of the WWTPs and the degradation mechanisms occurring in the water bodies. The model is designed to support decisions on the allocation and choice of additional treatment steps on a regional level. Our criterion for the assessment of beneficial effects is not the load reduction but the impact on micropollutants concentrations in the entire river network. Field sampling data have been used to calibrate the model for relevant pollutants like diclofenac and carbamazepine, and to verify its accuracy.

The model not only predicts spatially resolved exposure concentrations for micropollutants, identifying river sites with elevated concentrations, but also visualizes the impact of adopted measures *a priori* for compounds that are relevant for the Sûre river. This allows decision-makers to evaluate WWTP upgrade scenarios without the time- and cost-intensive measurement campaigns.

MATERIALS AND METHODS

Selection of target compounds

The current study monitors 27 micropollutants of different uses: 14 pharmaceuticals from 6 therapeutic classes known to be excreted in the highest amount in the Sûre catchment (i.e. antibiotics, beta-blockers, anti-inflammatories), 9 herbicides of emerging concern (i.e. glyphosate and its degradation product, AMPA) or with a legal obligation (i.e. carbendazim, diuron, and isoproturon), 2 fluorosurfactants with low EQS (i.e. perfluorooctanesulfonate (PFOS)) and other compounds known to be especially relevant for the Sûre river (i.e. benzotriazole and tris(2-chloroisopropyl)phosphate) (Gallé *et al.* 2019). These compounds are listed (Table 1) according to their CAS number and EQS values. Mean daily loads were calculated for nine selected substances for all river segments represented in the model. They are marked as (M) in the table.

Experimental unit set-up

Characteristics of the Sûre catchment

With its 173 km length, the Sûre river crosses three countries, rising from the Ardennes (Belgium), being the physical border between Germany and Luxembourg before emptying into the Mosel. The catchment is characterized by rural areas with municipalities connected to small and medium-sized WWTPs (below 50,000 PE).

Location of the pilot: the WWTP of Echternach

A CW-pilot plant was installed at the WWTP of Echternach (Luxembourg). This WWTP was chosen because of its good effluent quality (i.e. macropollutant values complying with the national legislation for limits of discharge) and its cross-border character treating 12,500 population equivalents (PE) from Germany out of 36,000 PE capacity, adequately representing

Table 1 | List of compounds

Compound	CAS number	Therapeutic Group/Use	AA-EQS[a] Chronic quality standard (μg L^{-1})
Pharmaceuticals			
Atenolol	29122-68-7	Beta-Blocker	150
Bezafibrate	41859-67-0	Lipid regulator	2.3
Carbamazepine (M)	298-46-4	Psychiatric drug	2
Clarithromycin (M)	81103-11-9	Antibiotic	0.12
Ciprofloxacin	85721-33-1	Antibiotic	0.089
Cyclophosphamide	50-18-0	Cytostatic	NA
Diclofenac (M)	15307-86-5	Anti-inflammatory	0.05
Erythromycin A	114-07-8	Antibiotic	NA
Ketoprofen	22071-15-4	Anti-inflammatory	NA
Lidocaine	137-58-6	Anesthetic	NA
Metoprolol	51384-51-1	Beta-Blocker	8.6
Propranolol	525-66-6	Beta-Blocker	0.16
N4-acetylsulfamethoxazole	21312-10-7	Metabolite	NA
Sulfamethoxazole (M)	723-46-6	Antibiotic	0.6
Pesticides/Herbicides etc.			
Carbendazim	10605-21-7	Fungicide	0.44
DEET	134-62-3	Insect repellent	88
Diuron	330-54-1	Algaecide	0.07
Isoproturon	34123-59-6	Algaecide	0.64
Terbutryn	886-50-0	Herbicide	0.065
Mecoprop (MCPP)	7085-19-0	Algaecide	3.6
Tolyltriazole (M)	29385-43-1	Fertilizer	NA
Glyphosate	1071-83-6	Herbicide	120
Aminomethylphosphonic acid (AMPA)	1066-51-9	Degradation product	1500
Fluorosurfactants			
Perfluorooctanesulfonic acid (PFOS) (M)	1763-23-1	Surfactant	0.002
Perfluorooctanoic acid (PFOA)	335-67-1	Surfactant	NA
Others			
Benzotriazole (M)	95-14-7	Corrosion inhibitor	240
Tris(2-chloroisopropyl)phosphate (TCPP) (M)	13674-84-5	Flame retardant	NA

[a]https://www.ecotoxcentre.ch/expert-service/quality-standards/proposals-for-acute-and-chronic-quality-standards/.

the catchment. The conventional activated sludge system consists of a primary clarifier (V=ca. 380 m^3) and two aerated reactors (V=ca. 4,500 m^3 each) integrated with secondary sedimentation (V=ca. 3,800 m^3), for a total Sludge Retention Time (SRT) of 19 d and a Hydraulic Retention Time (HRT) that varies between 26 and 31 h. The plant made also use of the grid, sand, and grease trap preceding the primary clarifier.

Description of the VF and sampling design

Two CWs (VF1 and VF2) of subsurface vertical flow configuration at pilot scale were evaluated as post-treatment steps for the removal of micropollutants. VF1 and VF2 exhibit a surface area of 11.18 m^2 and 12.78 m^2 respectively and were filled with a mix of sand (grain size 0–3 mm, Liapor, Germany) and 15% activated biochar as supporting material (grain size 2–5 mm, Palaterra, Germany), previously demonstrated to be suitable for this application (Venditti *et al.* 2022).

The systems were planted with macrophytes typical of a CW environment *Phragmites australis*, *Lythrum salicaria* and *Iris pseudacorus* (Brunhoferova *et al.* 2021) at a density of 25 plants steams per m². The effluent of the WWTP was used as influent to the units, pumped, and uniformly distributed over the surface. At each loading in the intermittent regime, the wastewater flooded the wetland surface, percolated by gravity through the wetland body, and collected in a 50 L plastic tank placed outside each unit. The feeding strategy generally consisted of six short equally daily water cycles (every 4 h) and the applied Hydraulic Loading Rate (HLR) varied from 100 to 200 L m^{-2} d^{-1}.

Several sampling campaigns took place over 14 months of operation to:

– characterize the contribution of the industrial (I) and the cross-border streams (both Luxembourgish, LU and German, DEU) into the influent: 24 h composite samples of each stream were collected in a time-proportional regime during two measurement campaigns;
– determine the occurrence of micropollutants and assess the efficiency of the conventional activated sludge treatment step: the raw wastewater (IN) was sampled after the mixing of the three streams and before entering the primary clarifier while effluent samples (EF) were collected after the secondary sedimentation. Four regular campaigns of 24 h composite samples were carried out. Following the guidelines of Koms (2021), two extended campaigns were additionally performed collecting 72 hrs composite samples for IN and EF samples. In both regular and extended campaigns, a delay of 24 hrs between IN and EF was considered to reflect the WWTP's HRT.
– assess the feasibility of VF wetland as post-treatment: grab samples (VF1 and VF2) were taken from 50 L volume after the units and related to the effluent of the WWTP.

A schematic of the sampling is presented in the Supplementary Information.

Analytical methodology

Macropollutants. Common parameters were routinely monitored. Chemical oxygen demand (COD), total nitrogen (TN), PO$_4$-P, NH$_4$-N, and NO$_3$-N were measured with Hach Lange cuvette tests. Oxidation-reduction potential, pH, and conductivity were collected with conventional WTW (Xylem, UK) Sensors.

Micropollutants. The analyses of the pharmaceuticals were performed externally (Luxembourg Institute of Science and Technology LIST, Luxembourg) and the methodology has been previously described (Venditti *et al.* 2022).

Calculation methods

To compare the contribution of industrial, Luxembourgish, and German wastewater streams into the WWTP inlet, the mass loads (MLs in g d^{-1} PE^{-1}) of relevant compounds have been calculated according to the following equation:

$$ML = \frac{C*Q}{PE}$$

where C is the measured pollutant concentration in ng L^{-1}, Q is the daily flow in L d^{-1}.

Per capita specific loads of 60 g Biological Oxygen Demand –5 days (BOD5), 120 g COD, and 11 g Total Kjeldahl Nitrogen (TKN) per capita and day are assumed.

To determine the efficiency of each treatment step, the elimination (E in %) of each compound has been calculated as:

$$E = \frac{(C_0 - C)}{C_0}*100$$

where C is the effluent and C$_0$ is the influent concentration. The eliminations were calculated for each campaign and then an average was considered.

Modelling relevant compounds with the EmiSûre model

General model approach

In the EmiSûre project, a mass balance model tailored for the application in the Greater Region was developed. The model calculates geo-referenced concentrations in surface waters taking into account point emissions from WWTPs and Combined

Sewer Overflows (CSO). Concentrations are simulated in a stationary way for mean minimal and mean annual flow conditions separately. Emissions of WWTPs are expressed as average substance-specific loads (L(S) in g d^{-1}). To calculate these point emissions, the model uses the PE connected to each WWTP (P$_{WWTP}$ in I) and per-capita loads (l(s) in g I^{-1} d^{-1}) in the influent of the WWTPs for each considered substance S. Wastewater treatment is modeled as a constant removal process whose removal efficiencies are depending on the type of applied treatment (conventional WWTP f$_{WWTP}$, advanced treatment process f$_{ATP}$ in %) and the considered substance S.

To estimate the influence of CSO events on surface water quality, a CSO discharge factor (f(S)$_{CSO}$ in %) was established. It expresses the proportion of a substance which enters the surface waters by CSO discharge and thus does not pass the WWTP.

The general model approach is illustrated in Figure 1.

The surface water network is represented by river segments between nodes i. The nodes are defined by load and concentration changes in the river (e.g. point emissions). Each segment is assigned to a flow rate (e.g. mean annual flow) and the corresponding flow velocity. The loads are calculated for each river segment based on a steady-state mass-balance approach. Emitted loads are introduced at the beginning of each river segment (L(S)$_i$). The loads at the end of each river segment (L(S)$_{i+1}$) are calculated taking into account the transformation of the compounds: a first-order reaction kinetic is assumed using a loss constant (k(S) in h^{-1}) specific for each substance and the hydraulic retention time (HRT in s) in the river segment.

Assuming constant emissions (mass per time) from wastewater systems, the calculated loads represent average annual conditions (Knerr *et al.* 2020).

Model setup

A total river system length of 899 km (which includes Sûre and all tributaries) was represented and divided into 568 individual segments. The segment-specific flow was included based on historical time series from 29 gauges in the catchment. Information on the range of flow conditions was included by implementing mean annual flow and mean minimal annual flow conditions. The model includes all 286 WWTPs of the catchment and the associated CSOs. All drainage systems in the catchment were considered as a combined system, since 97% of the sewers in the catchment represent combined sewer systems. Uniform input data concerning substance-specific loads have been applied all over the catchment. Data for substance-specific loads, removal efficiencies in WWTPs, etc. were taken from literature as mean values. Model input data and used literature are listed in the Supplementary Information.

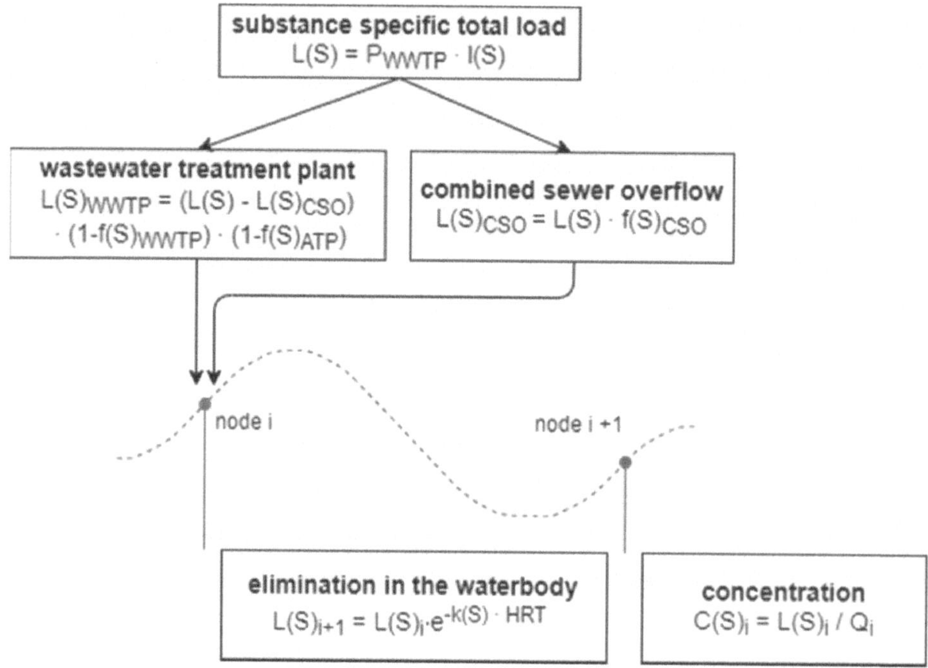

Figure 1 | General model approach.

Assessment of receiving water concentrations

If model predictions for pollutant loads do not significantly deviate from calculated loads using measured data, a Predicted Environmental Concentration (PEC(S)) for the substance S can be calculated. To this end, daily loads ($L(S)_i$ in kg d^{-1}) were divided by daily flows (Q_i in m^3 d^{-1}) in the river. Mean annual flow conditions were used to account for average concentrations, mean minimal flow conditions were used to account for minimum dilution and accordingly maximum concentrations in the waters. Emissions from CSOs have been included in the simulation only for mean annual flow.

Pollutant concentrations were assessed based on the so-called Contamination Factor (CF). This is the ratio of the calculated water concentrations (PEC(S)) and Water Quality Criteria (WQC(S)) defined for each substance. As WQC, concentrations excluding chronic damage or short-term toxicity of aquatic organisms were chosen. For Water Quality Standard (WQS), values defined in the Directive 2013/39/EU (EC 2013) and in the German water surface act (OGewV 2016) were used. Annual Average Environmental Quality Standards (AA-EQS) were combined with mean flow conditions and Maximal Acceptable Concentrations (MAC-EQS) were combined with minimal flow conditions. Most of the substances identified in the Sûre river catchment are currently not regulated legally. For this reason, replacement values taken from the literature (e.g. Predicted No Effect Concentration (PNEC(S)) was used. The results are displayed in the form of color-coded maps for mean annual and mean minimal annual flow conditions in the river system. Hence, river segments with CF >1, which means concentrations higher than WQC(S), can easily be identified.

Minimizing emission: strategies

To avoid any exceedance of a WQS, WWTPs will be upgraded with additional treatment steps in the model regardless of their design capacity. To examine the influence of the WWTP upgrade, the simulation starts at the first WWTP in a river where the WQC is exceeded downstream. The effect of the WWTP upgrade is then evaluated step by step downstream, taking into account the capacity of the rivers. The procedure is repeated for all rivers. For WWTPs with a design capacity >10,000 PE, ozonation or activated carbon filtration were chosen as post-treatment technology. For WWTPs with a design capacity of <10,000 PE, constructed wetlands were applied with the experimental results of EmiSûre as input data. To minimize water concentrations, other measures such as source-oriented measures (consumption reduction, substitution, etc.) can be considered.

RESULTS AND DISCUSSION

Feasibility of VF for enhanced removal strategy

Occurrence and removal efficiencies in the WWTP of Echternach

All the streams contributing to the WWTP's influent were characterized in terms of their physicochemical parameters (Table 2). The German daily flow resulted in being almost 2 and 4 times the industrial and Luxembourgish streams, respectively. High BOD and COD values suggested wastewater polluted with nonbiodegradable contaminants originating from

Table 2 | General parameters and macropollutants concentrations in 24 h composite samples, $N = 6$

Parameter	I	LU	DEU
Q (m^3 d^{-1})	628	1,510	2,560
PE BOD	10,571	2,869	12,971
COD (mg L^{-1})	1,780	200	545
BOD (mg L^{-1})	1,010	114	304
TN (mg L^{-1})	26	27	44
Ammonia (mg L^{-1})	8.5	17	28
Ptot (mg L^{-1})	9.3	3.2	4.2
Na (mg L^{-1})	279	60	53
K (mg L^{-1})	76	12	16
Ca (mg L^{-1})	53	100	82
Conductivity (µS cm^{-1})	1,699	1,068	1,008

anthropogenic activities driven by the local industry (automotive and composite materials factories, etc). Likewise, relevant concentrations of Na, K, Ca, and conductivity as indicators of high dissolved solids or minerals were also observed in the industrial stream.

The presence of micropollutants in the WWTP's influent was evaluated by tracking back to their origin within the three streams. Micropollutants can get into surface water also out of emissions from industrial processes, which may use chemicals as adjuvants in their operations. Chemicals adjuvants include complexing agents, surfactants, and preservatives that are not usually used in private households.

When looking at the daily unit mass loads per inhabitant (Table 3) calculated per PE, only 7 substances out of the 27 monitored micropollutants were detected in non-relevant concentrations:

- the antifungal carbendazim together with diuron, isoproturon, terbutryn (Bollmann *et al.* 2014), and MCPP registered as herbicides, forbidden in agriculture but still used as biocides or fungicides in roof paints, outside wall paints, and coating. Their content below the Limit of Quantification (LOQ) indicates that WWTP is not a relevant source of emission. The presence of these contaminants in some European rivers can thus be explained by the diffuse pollution from stormwater run-off (Quednow & Püttmann 2009);

Table 3 | Mass loads of 20 relevant compounds (g d^{-1} PE^{-1}) in 24 h composite samples for industrial, Luxembourgish, and German wastewater streams, $N = 6$

Parameter	I (g d^{-1} PE^{-1})	LU (g d^{-1} PE^{-1})	DEU (g d^{-1} PE^{-1})
Atenolol	0.0013	0.4750	0.1697
Bezafibrate	0.0179	0.0032	0.0081
Carbamazepine	0.0009	0.2916	0.1995
Clarithromycin	0.0009	0.2916	0.1995
Ciprofloxacin	0.0003	0.3747	0.0408
Cyclophosphamide	<LOQ	<LOQ	<LOQ
Diclofenac	0.0234	1.4847	0.7102
Erythromycin	<LOQ	<LOQ	<LOQ
Ketoprofen	0.0002	0.1388	0.0358
Lidocaine	0.0002	0.0322	0.0205
Metoprolol	0.0289	0.0496	0.2211
Propranolol	0.0008	0.1475	0.0056
N-acetyl sulfamethoxazole	0.0003	0.3296	0.2645
Sulfamethoxazole	0.0004	0.1347	0.1171
Benzotriazole	0.0179	0.0032	0.0081
Carbendazim	<LOQ	<LOQ	<LOQ
DEET	0.0077	0.0979	0.0781
Diuron	<LOQ	<LOQ	<LOQ
Isoproturon	<LOQ	<LOQ	<LOQ
Terbutryn	<LOQ	<LOQ	<LOQ
MCPP	<LOQ	<LOQ	<LOQ
TCPP	0.1509	2.3187	3.4719
Tolyltriazole	0.0280	1.0229	0.4078
Glyphosate	0.0044	0.0816	0.0730
AMPA	1.0337	0.2211	0.1358
PFOS	0.0133	0.0101	0.0067
PFOA	0.0050	0.0081	0.0069

- the cytostatic cyclophosphamide, usually administrated in low amount if compared to other medicaments of this type (i.e. 5-fluorouracil). Only 10% of the administrated active ingredient is excreted unchanged via urine, while metabolites and transformation products are not considered in this study;
- the antibiotic erythromycin: among macrolides, this antibiotic mostly suffers from analytical detection problems. Also, only 5% of the administrated active ingredient is excreted unchanged via urine.

For the remaining 20 relevant compounds, the industrial stream resulted in being the main source of AMPA, bezafibrate, and PFOS while micropollutants of medical use appeared minimal (i.e. atenolol, carbamazepine, diclofenac, etc). On the contrary, both Luxembourgish and German domestic wastewater streams significantly contribute to micropollutants like benzotriazole, diclofenac, TCPP, and tolytriazole. As AMPA usually co-occurs with glyphosate, their transport is traditionally described together. However, previous researchers (Grandcoin *et al.* 2017) indicated industrial phosphonate chelating agents as an alternative source of AMPA, which is in line with the strong contribution of the industrial stream identified in this study. It is interesting to observe that among beta-blockers, atenolol and propranolol seemed to be more administrated in Luxembourg, with mass loads 3 and 26 times higher than in Germany, respectively. On the contrary, metoprolol was four times higher in Germany than in Luxembourg.

The concentration of the relevant substances was also measured to assess the removal in conventional WWTP treatment steps (Table 4).

For a few compounds, the effluent concentrations were higher than the influent ones. The reasons for that could be linked to the experimental methodology (i.e. analytical and sampling) or the fate mechanism of the single compound:

- carbamazepine and lidocaine are known to be persistent and hardly removed (Falås *et al.* 2016; Gallé *et al.* 2019) in Conventional Activated Sludge (CAS) treatments. Additionally, carbamazepine metabolites can build back to the parent compound (Ternes 1998; Bahlmann *et al.* 2014; Scheurer *et al.* 2015). A zero removal is thus assumed;
- the acetyl functional group of N-acetyl sulfamethoxazole breaks during the wastewater treatment process, generating the mother compound (sulfamethoxazole) (Göbel *et al.* 2005);

Table 4 | Average concentrations (Min-Max) measured during the observation period (14 months) for 20 relevant compounds (ng L^{-1}) and WWTP removal efficiencies (%), $N=12$

Compound	IN (ng L^{-1})	EF (ng L^{-1})	Elimination (%)
Atenolol	323 (65–680)	43 (24–65)	70
Bezafibrate	130 (57–268)	29 (8–60)	73
Carbamazepine	578 (129–1,262)	529 (96–1,290)	0
Clarithromycin	504 (64–1,015)	276 (61–583)	33
Ciprofloxacin	654 (145–3,229)	218 (147–357)	37
Diclofenac	2,109 (302–4,070)	1,523 (481–2,508)	<20
Ketoprofen	111 (15–211)	10 (4–29)	85
Lidocaine	92 (20–151)	122 (32–231)	0
Metoprolol	524 (175–1,150)	363 (202–720)	<10
Propranolol	167 (27–352)	131 (31–265)	<10
N-acetyl sulfamethoxazole	244 (33–658)	10 (4–28)	93
Sulfamethoxazole	165 (4–562)	79 (18–176)	−34
Benzotriazole	4,905 (1,180–10,285)	2,337 (1,049–3,493)	41
DEET	371 (46–1,845)	40 (19–59)	75
TCPP	3,151 (1,320–5,851)	3,002 (1,076–5,218)	<10
Tolyltriazole	1,487 (590–2,602)	1,000 (505–1,337)	<20
Glyphosate	100 (23–196)	127 (60–193)	−27
AMPA	2,341 (825–4,650)	4,079 (1,500–9,460)	−74
PFOS	120 (10–413)	61 (5–246)	22
PFOA	20 (5–50)	15 (4–47)	0

- AMPA is a degradation product of glyphosate and/or phosphonates. As such, its fate has to be related to the parent compounds (Grandcoin *et al.* 2017).

Benzotriazole, TCPP, AMPA, diclofenac, and tolyltriazole were detected as the most abundant in the WWTP influent, exceeding 1,000 ng L^{-1} average concentrations. Among them, TCPP, diclofenac, and tolyltriazole were poorly removed while benzotriazole showed a moderate removal together with the antibiotics ciprofloxacin and clarithromycin.

Performance of VF

The intermittent regime was used as the basis for the operation of VF1 and VF2 because it is known to improve the oxidation-reduction conditions favorable for complete nitrification and elimination of micropollutants. Clogging of the medium has also been demonstrated to be less frequent if compared with continuous flow operation. Among 27 compounds, only 15 were considered relevant in the effluent of the WWTP as influent to VF1 and VF2. An average concentration and removal were considered for the post-treatment step.

Results show (Table 5) high removal efficiencies for most relevant compounds with the only exception of AMPA and TCPP, which are still discharged in relevant concentrations. These results confirm those already achieved in a previous pilot installed in a WWTP with less than 10,000 PE, operated for 6 months (Venditti *et al.* 2022).

When relative contributions of the two treatment steps are considered (conventional WWTP and post-treatment with VF, Figure 2) to the overall elimination of the selected compounds, the VF results are crucial in the elimination of persistent compounds (i.e. carbamazepine and lidocaine), antibiotics (especially clarithromycin and ciprofloxacin), diclofenac, benzotriazole, and tolyltriazole. The overall removal of TCPP is, however, still poor. The use of a VF for the WWTP of Echternach allows complying with the 80% removal threshold for the four mandatory compounds defined by the Luxembourgish Water Administration, diclofenac, carbamazepine, clarithromycin, and benzotriazole (Administration de Gestion de l'eau 2020), with the average removal rate of 99, 99, 97, and 100%, respectively.

Modeling loads and concentrations of micropollutants in the Sûre

Mean annual loads

Mean daily loads were calculated for nine selected substances for all river segments represented in the model. The selection followed the methodology proposed in Knerr *et al.* (2020) and represents the substances that can be predicted with the model

Table 5 | Average concentrations (Min-Max) measured during the observation period (14 months) for 15 relevant compounds (ng L^{-1}) and VF removal efficiencies (%), $N=12$

Compound	IN VF (ng L^{-1})	EF VF (ng L^{-1})	Elimination VF (%)
Carbamazepine	529 (96–1,290)	6 (4–13)	98
Clarithromycin	276 (61–583)	14 (3–43)	95
Ciprofloxacin	218 (147–357)	36 (5–178)	92
Diclofenac	1,523 (481–2,508)	11 (4–31)	99
Lidocaine	122 (32–231)	4 (3–20)	97
Metoprolol	363 (202–720)	24 (5–80)	93
Propranolol	131 (31–265)	13 (5–42)	88
N-acetyl sulfamethoxazole	10 (4–28)	4 (3–5)	45
Sulfamethoxazole	79 (18–176)	5 (4–7)	89
Benzotriazole	2,337 (1,049–3,493)	17 (8–39)	99
TCPP	3,002 (1,076–5,218)	2,075 (707–5,761)	26
Tolyltriazole	1,000 (505–1,337)	11 (4–41)	99
Glyphosate	127 (60–193)	19 (5–49)	85
AMPA	4,079 (1,500–9,460)	2,634 (66–5,720)	25
PFOS	61 (5–246)	19 (5–67)	62

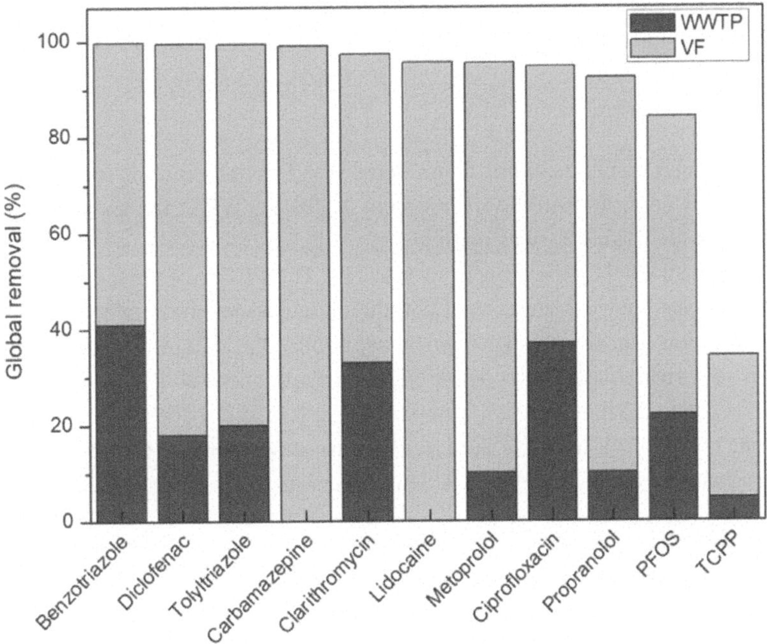

Figure 2 | Relative contributions of the VF and the WWTP to the overall global average removal efficiencies.

approach reliably. To evaluate the validity of the setup, predicted loads were compared with loads calculated from measurements. For this purpose, data from different monitoring campaigns carried out by the local authorities were used. Additionally, a measuring campaign in the rivers of the Sûre for the selected micropollutants (see Table 1) was carried out in this study.

The comparison of predicted vs calculated daily loads for the pharmaceuticals diclofenac and carbamazepine is plotted in Figure 3. The results show that the model reflects the loads in the rivers very well for these two substances, because of their constant consumption in the catchment throughout the year and in all regions. The maximal deviation for these two substances was around a factor of two and therefore in the same range as in comparable studies (Ort *et al.* 2009; Alder *et al.* 2010).

Similarly, good agreements were achieved for benzotriazole, clarithromycin, sulfamethoxazole, TCPP, PFOS, and tolyltriazole. However, for these substances predicted loads were evaluated against a reduced number of measurements because these substances were often not detectable in the river samples (Supplementary Information).

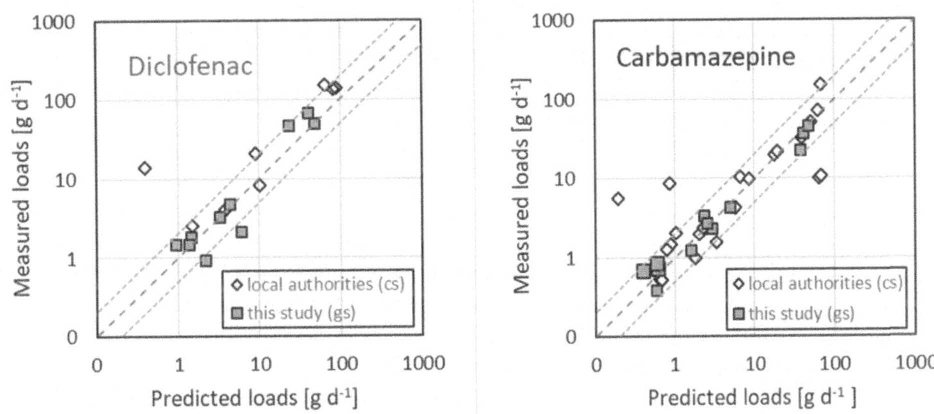

Figure 3 | Model prediction vs measured daily diclofenac and carbamazepine loads (cs=composite samples, gs = grab samples).

Metoprolol is overestimated at Luxembourg monitoring sites, while at German monitoring sites a good agreement was achieved. This is probably due to variations in regional differences in prescribing behavior for this drug (Supplementary Information).

Water contamination

Figure 4 illustrates the Sûre River catchment including all modeled WWTPs (left) and the status quo for diclofenac in the form of a CF-map for mean annual flow conditions in the river system (right). As WQC, the proposed AA-EQS, recommended by the Ecotox center of Switzerland (see Table 1), was applied.

CF >0 is obtained in all waters affected by the effluent of WWTPs. The PEC exceeds the proposed AA-EQS for diclofenac of 0.05 µg L^{-1} in.14% of the represented river segments. High diclofenac concentrations with CF >2 are partly encountered in upper reaches due to low dilution (e.g. river Wiltz, river Nims) and in particular in rivers, which are highly loaded by the discharge of micropollutants (e.g. river Alzette) due to the higher urban connections.

In the Sûre river itself, CF is below 0.5 in about 82% and 0.5≤CF<1.0 in about 16% of the total river length. However, this does not imply that diclofenac concentrations in the Sûre river always stay below the estimated WQC of 0.05 µg L^{-1}. Besides diclofenac, only PFOS exceedance of the estimated WQC was observed in the model results.

Immission-based reduction strategy

The model was used to develop strategies that avoid any exceedance of WQS. We focused on diclofenac because compared to PFOS, diclofenac is eliminated with high efficiency in post-treatment steps. Thus, diclofenac represents a sensitive parameter for the selection of WWTPs to be upgraded.

The result of the applied immission-based reduction strategy for diclofenac is plotted in Figure 5 (right). The upgraded WWTPs are highlighted in Figure 5 (left). Elimination of 94% is assumed for ozonation and 83% for activated carbon filters corresponding to the treatment efficiency found in the literature) (Abegglen & Siegrist 2012; Margot *et al.* 2013; Götz *et al.* 2015). For constructed wetlands, an elimination of 99% was used, corresponding to the findings of this study.

An upgrade of smaller WWTPs in the upper reaches of the river is necessary to avoid any exceedance of WQC. This has also a positive impact on the downstream water bodies and additionally on the evaluation of WWTPs further downstream. To

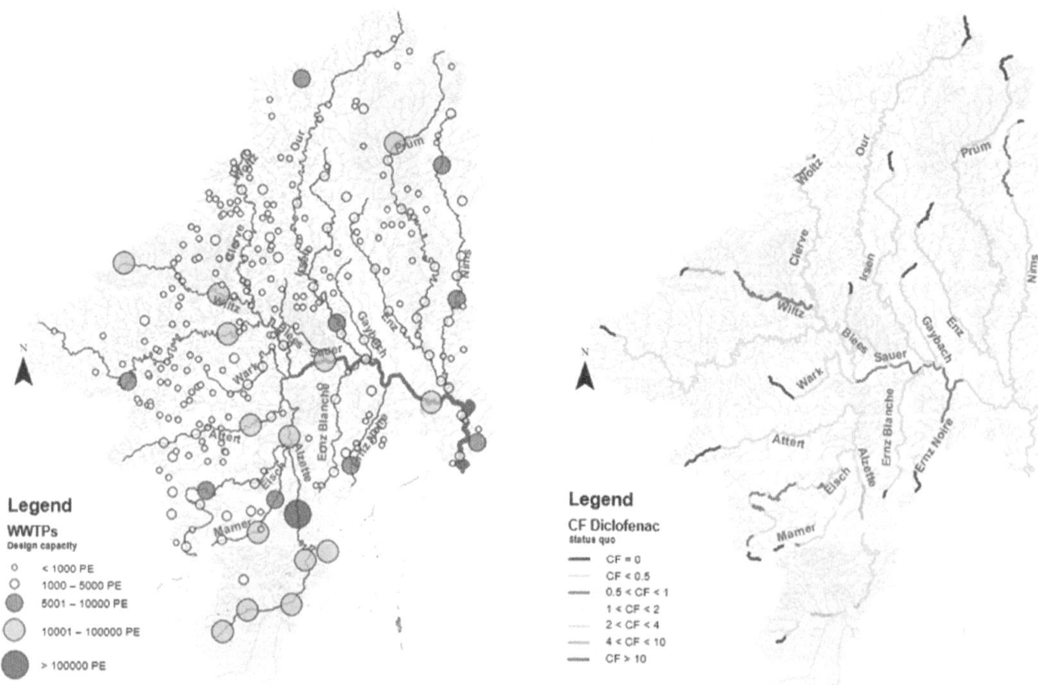

Figure 4 | Sûre River catchment including all modeled WWTPs (left) and the ratio of predicted diclofenac concentrations and proposed AA-EQS for mean annual flow (right).

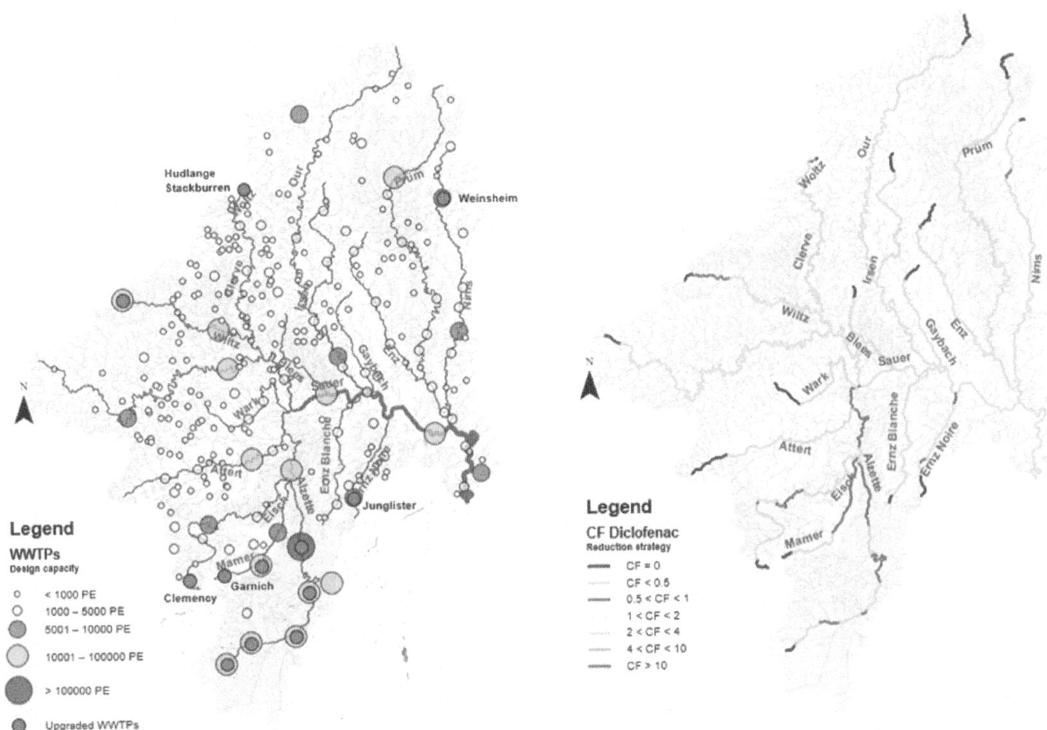

Figure 5 | Upgraded WWTPs (left) and the ratio of predicted diclofenac concentrations and proposed AA-EQS for mean annual flow for the emission-based reduction strategy (right).

avoid the exceedance of the proposed AA-EQS for diclofenac, a minimum of 12 WWTPs would have to be upgraded. This includes one WWTP with a design capacity of less than 10,000 PE (Junglister/LU) and four WWTPs with a design capacity of less than 5,000 PE (Garnich, Huldange Stackburren, Clemency in LU and Weinsheim in DEU).

The selection of WWTPs to be upgraded is based on the resulting concentrations for diclofenac downstream of each WWTP at mean annual flow conditions. Furthermore, diclofenac allows evaluation of the quality-related benefit in the river catchment via the additional kilometers with CF <1 (Table 6). PFOS as the second parameter with exceedances in the current state cannot be used for this purpose, since PFOS shows an inefficient elimination by ozonation (5%) and activated carbon filters (55%) (Pinnekamp & Markel 2008).

However, other factors are also important for the overall assessment of measures, such as the load reduction at the outlet of the catchment. Table 6 describes the resulting load reduction at the mouth of the Sûre river (load-related) and the portion of the water system with CF <1 (quality-related). Even if substances such as benzotriazole or tolyltriazole are not relevant in

Table 6 | Load- and quality-related comparison at mean annual flow conditions (WQS are exceeded only for diclofenac and PFOS)

Substance	Quality-related benefit additional km with CF <1 (km)	Load-related benefit resulting in overall load reduction (g d^{-1})
Benzotriazole	0	−498
Carbamazepine	0	−29
Clarithromycin	0	−17
Diclofenac	118	−21
Sulfamethoxazole	0	−4.8
Tolyltriazole	0	−165
PFOS	0	−0.1
TCPP	0	−38

terms of a quality-related evaluation, the upgrade of WWTPs results in a significant load reduction in the catchment and thus also in the downstream waters. The total load reduction potential is significantly higher, since only a small number of substances found in the rivers were modeled.

CONCLUSIONS, RECOMMENDATIONS, AND PERSPECTIVE

Implementing water protection policies for catchments in rural areas generally requires the know-how of different specialists from the identification of relevant pollutants to the subsequent implementation of measures. In this study, we attempted to give the means to decision-makers of the Greater Region with dominant rural areas and beyond.

Considering the Sûre catchment representative, a mass balance model has been developed, successfully used, and evaluated with measured field concentrations to first assess the status quo of the catchment and then predict concentrations of relevant compounds (i.e. diclofenac and PFOS in this case). The developed model is a simplified approach for decision making on upgrading of WWTPs that neglects dynamics of CSOs. It should not transfer to substances that mainly originate from surface runoff.

In parallel, CWs in VF configuration have been tested and resulted in being efficient in the removal of micropollutants complying with local regulations for discharge. The experimental results in terms of removal efficiencies for the relevant compounds have been used as input data for the EmiSûre model when a catchment management scenario was utilized such as upgrading WWTP >10,000 PE with activated carbon filtration and ozonation, and CWs in VF configuration for WWTP <10,000 PE. When looking at the effect of the measure, the selection of the WWTPs to be upgraded resulted in a significant load reduction in the catchment and thus in the downstream waters.

ACKNOWLEDGEMENTS

The presented outcomes are part of the European EmiSûre project (n 013-2-03-049), which is co-founded by the EU INTERREG VA program. The authors are especially thankful to: the Administration de la Gestion de l'Eau (AGE) located at the Ministère de l'Intérieur et de l'Aménagement du Territoire in Luxembourg and Ministerium für Wirtschaft, Verkehr, Landwirtschaft und Weinbau Rheinland-Pfalz (MWVLW) in Germany.

DATA AVAILABILITY STATEMENT

All relevant data are included in the paper or its Supplementary Information.

CONFLICT OF INTEREST STATEMENT

The authors declare there is no conflict.

REFERENCES

Abegglen, C. & Siegrist, H. 2012 Mikroverunreinigungen aus kommunalem Abwasser. Verfahren zur weitergehenden Elimination auf Kläranlagen. *Umwelt-Wissen* **1214** (210 S). Available from: https://www.dora.lib4ri.ch/eawag/islandora/object/eawag%3A14520.

Administration de Gestion de l'eau. 2020 9. *Recommandations Regardant la Mise en œuvre D'une Quatrième étape épuratoire sur les Stations D'épuration Municipales.*

Alder, A. C., Schaffner, C., Majewsky, M., Klasmeier, J. & Fenner, K. 2010 Fate of β-blocker human pharmaceuticals in surface water: comparison of measured and simulated concentrations in the Glatt Valley Watershed, Switzerland. *Water Research* **44** (3). https://doi.org/10.1016/j.watres.2009.10.002.

Bahlmann, A., Brack, W., Schneider, R. J. & Krauss, M. 2014 Carbamazepine and its metabolites in wastewater: analytical pitfalls and occurrence in Germany and Portugal. *Water Research* **57**. https://doi.org/10.1016/j.watres.2014.03.022.

Bollmann, U. E., Vollertsen, J., Carmeliet, J. & Bester, K. 2014 Dynamics of biocide emissions from buildings in a suburban stormwater catchment – concentrations, mass loads and emission processes. *Water Research* **56**. https://doi.org/10.1016/j.watres.2014.02.033.

Brunhoferova, H., Venditti, S., Schlienz, M. & Hansen, J. 2021 Removal of 27 micropollutants by selected wetland macrophytes in hydroponic conditions. *Chemosphere* **281**. https://doi.org/10.1016/j.chemosphere.2021.130980.

EC 2013 *Directive 2013/39/EU of 12 August 2013 Amending Directives 2000/60/EC and 2008/105/EC as Regards Priority Substances in the Field of Water Policy.*

Eggen, R. I. L., Hollender, J., Joss, A., Schärer, M. & Stamm, C. 2014 Reducing the discharge of micropollutants in the aquatic environment: the benefits of upgrading wastewater treatment plants. *Environmental Science and Technology* **48** (14). https://doi.org/10.1021/es500907n.

Falås, P., Wick, A., Castronovo, S., Habermacher, J., Ternes, T. A. & Joss, A. 2016 Tracing the limits of organic micropollutant removal in biological wastewater treatment. *Water Research* **95**. https://doi.org/10.1016/j.watres.2016.03.009.

Gallé, T., Koehler, C., Plattes, M., Pittois, D., Bayerle, M., Carafa, R., Christen, A. & Hansen, J. 2019 Large-scale determination of micropollutant elimination from municipal wastewater by passive sampling gives new insights in governing parameters and degradation patterns. *Water Research* **160**. https://doi.org/10.1016/j.watres.2019.05.009.

Göbel, A., Thomsen, A., McArdell, C. S., Joss, A. & Giger, W. 2005 Occurrence and sorption behavior of sulfonamides, macrolides, and trimethoprim in activated sludge treatment. *Environmental Science and Technology* **39** (11). https://doi.org/10.1021/es048550a.

Götz, C., Otto, J. & Singer, H. 2015 *Überprüfung des Reinigungseffekts. Aqua & Gas No. 2*, pp. 34–40.

Grandcoin, A., Piel, S. & Baurès, E. 2017 Aminomethylphosphonic acid (AMPA) in natural waters: its sources, behavior and environmental fate. *Water Research* **117**. https://doi.org/10.1016/j.watres.2017.03.055.

Ilyas, H. & van Hullebusch, E. D. 2019 Role of design and operational factors in the removal of pharmaceuticals by constructed wetlands. *Water (Switzerland)* **11** (11). https://doi.org/10.3390/w11112356.

Ilyas, H., Masih, I. & van Hullebusch, E. D. 2020 Pharmaceuticals' removal by constructed wetlands: a critical evaluation and meta-analysis on performance, risk reduction, and role of physicochemical properties on removal mechanisms. *Journal of Water and Health* **18** (3). https://doi.org/10.2166/wh.2020.213.

Knerr, H., Gretzschel, O., Valerius, B., Srednoselec, I., Zhou, J., Schmitt, T. G., Steinmetz, H., Dittmer, U., Taudien, Y. & Kolisch, G. 2020 Modellgestützte Bilanzierung von Mikroschadstoffen in Gewässern gwf Wasser Abwasser **161** (3), 55–65.

Koms 2021. Available from: http://www.koms-bw.de/klaeranlagen/uebersichtskarte/.

Li, Y., Zhu, G., Ng, W. J. & Tan, S. K. 2014 A review on removing pharmaceutical contaminants from wastewater by constructed wetlands: design, performance and mechanism. *Science of the Total Environment* **468–469**. https://doi.org/10.1016/j.scitotenv.2013.09.018.

Margot, J., Kienle, C., Magnet, A., Weil, M., Rossi, L., de Alencastro, L. F., Abegglen, C., Thonney, D., Chèvre, N., Schärer, M. & Barry, D. A. 2013 Treatment of micropollutants in municipal wastewater: ozone or powdered activated carbon? *Science of the Total Environment* **461–462**. https://doi.org/10.1016/j.scitotenv.2013.05.034.

OGewV 2016 *Verordnung zum Schutz der Oberflächengewässer*.

Ort, C., Hollender, J., Schaerer, M. & Siegrist, H. 2009 Model-based evaluation of reduction strategies for micropollutants from wastewater treatment plants in complex river networks. *Environmental Science and Technology* **43**, 9. https://doi.org/10.1021/es802286v.

Pinnekamp, J. & Markel, W. 2008 *Senkung des Anteils Organischer Spurenstoffe in der Ruhr Durch Zusätzliche Behandlungsstufen auf Kommunalen Kläranlagen – Güte und Kostenbetrachtungen*. Abschlussbericht Forschungsvorhaben, Ministerium für Umwelt und Naturschutz, Landwirtschaft und Verbraucherschutz des Landes Nordrhein-Westfalen, Aachen/Mülheim.

Quednow, K. & Püttmann, W. 2009 Temporal concentration changes of DEET, TCEP, terbutryn, and nonylphenols in freshwater streams of Hesse, Germany: possible influence of mandatory regulations and voluntary environmental agreements. *Environmental Science and Pollution Research* **16** (6). https://doi.org/10.1007/s11356-009-0169-6.

Scheurer, M., Heß, S., Lüddeke, F., Sacher, F., Güde, H., Löffler, H. & Gallert, C. 2015 Removal of micropollutants, facultative pathogenic and antibiotic resistant bacteria in a full-scale retention soil filter receiving combined sewer overflow. *Environmental Sciences: Processes and Impacts* **17** (1). https://doi.org/10.1039/c4em00494a.

Ternes, T. A. 1998 Occurrence of drugs in German sewage treatment plants and rivers. *Water Research* **32** (11). https://doi.org/10.1016/S0043-1354(98)00099-2.

UBA 2009 *Energieeffizienz kommunaler Kläranlagen, Umweltbundesamt, Forschungsbericht. UBA-FB 001075*(205 26 307).

Venditti, S., Brunhoferova, H. & Hansen, J. 2022 Behaviour of 27 selected emerging contaminants in vertical flow constructed wetlands as post-treatment for municipal wastewater. *Science of the Total Environment* **819**. https://doi.org/https://doi.org/10.1016/J.SCITOTENV.2022.153234.

First received 30 March 2022; accepted in revised form 6 June 2022. Available online 13 June 2022

doi: 10.2166/wst.2022.179

Comparison of simple models for total nitrogen removal from agricultural runoff in FWS wetlands

Alba Canet-Martí [a,*], Sabrina Grüner[a,b], Stevo Lavrnić [c], Attilio Toscano [c], Thilo Streck [b] and Guenter Langergraber [a]

[a] Department of Water, Atmosphere and Environment, Institute of Sanitary Engineering and Water Pollution Control, University of Natural Resources and Life Sciences, Vienna (BOKU), Muthgasse 18, Vienna 1190, Austria
[b] Department of Biogeophysics, University of Hohenheim, Emil-Wolff-Str. 27, Stuttgart 70599, Germany
[c] Department of Agricultural and Food Sciences, Alma Mater Studiorum-University of Bologna, Viale Giuseppe Fanin 50, Bologna 40127, Italy
*Corresponding author. E-mail: alba.canet@boku.ac.at

AC-M, 0000-0002-7791-5837; SL, 0000-0002-0773-4360; AT, 0000-0001-5477-0623; TS, 0000-0001-7822-7588; GL, 0000-0003-4334-9563

ABSTRACT

Free water surface (FWS) wetlands can be used to treat agricultural runoff, thereby reducing diffuse pollution. However, as these are highly dynamic systems, their design is still challenging. Complex models tend to require detailed information for calibration, which can only be obtained when the wetland is constructed. Hence simplified models are widely used for FWS wetlands design. The limitations of these models in full-scale FWS wetlands is that these systems often cope with stochastic events with different input concentrations. In our study, we compared different simple transport and degradation models for total nitrogen under steady- and unsteady-state conditions using information collected from a tracer experiment and data from two precipitation events from a full-scale FWS wetland. The tanks-in-series model proved to be robust for simulating solute transport, and the first-order degradation model with non-zero background concentration performed best for total nitrogen concentrations. However, the optimal background concentration changed from event to event. Thus, to use the model as a design tool, it is advisable to include an upper and lower background concentration to determine a range of wetland performance under different events. Models under steady- and unsteady-state conditions with simulated data showed good performance, demonstrating their potential for wetland design.

Key words: agricultural runoff, design models, free water surface wetlands, modelling, treatment wetlands

HIGHLIGHTS

- The most robust model combination: the tanks-in-series model under steady conditions.
- First-order kinetics with non-zero background concentration was the best for degradation.
- The optimal background concentration changed from event to event.
- Potential use of TIS model in unsteady conditions when infiltration and evapotranspiration occur.

GRAPHICAL ABSTRACT

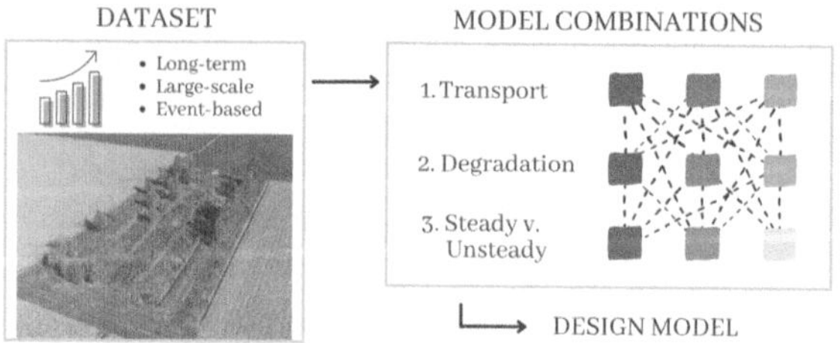

INTRODUCTION

The loss of water-soluble nutrients from agricultural soils through runoff and percolation has made agricultural systems one of the main sources of diffuse pollution of water bodies in recent decades (Robertson & Vitousek 2009; Withers *et al.* 2014). One of the most important pollutant of that origin is nitrogen that can cause negative effects in different environmental matrices, and the progressive increase of nitrates in soils and freshwaters has been recognised as a serious problem (Mancuso *et al.* 2021).

Among aquatic systems, wetland systems are known for their high productivity, as they can transform and accumulate organic matter, nutrients and other incoming compounds (Brinson *et al.* 1981). This results in a high purification capacity, which has led to the development of treatment wetlands (TWs), a nature-based solution used to treat different types of wastewater, including agricultural runoff (Langergraber *et al.* 2019). The most widely used TW type for agricultural runoff are free-water surface flow (FWS) wetlands due to their ability to cope with stochastic runoff events. FWS wetlands require large surface areas that are exposed to precipitation and evaporation. This influences the water balance, e.g. causing the system to dry out if rainfall does not compensate for evapotranspiration. In addition, they usually have a permeable bottom, hence infiltration water losses. Their performance depends on different factors, such as surface area, hydraulic residence time (HRT), hydraulic loading rate (HLR) or ambient temperature, which makes their design more challenging (Kadlec & Wallace 2009; Dotro *et al.* 2017).

The growing interest in FWS wetlands in the 1980s led to the development of mathematical models to predict their behaviour. Since then, various models have been used for design or optimization purposes, ranging from the simplest models, based on empirical equations that establish a relationship between input and output, i.e. linear regression (Kadlec 1994; Stone *et al.* 2004; Kadlec & Wallace 2009), to process-based models (Hammer & Kadlec 1986; Gargallo *et al.* 2017; Aboukila & Deng 2018; Aragones *et al.* 2020). The latter require modelling skills and several parameters to be calibrated, which can only be obtained when the wetland is constructed. In addition, when there is insufficient information for calibration, the complex models tend to propagate uncertainty errors, making them cumbersome and expensive to use for decision-making and upfront design. On the other hand, simple tools can provide an overview of the performance of TWs regarding water quality under different conditions, support their design and increase confidence in their implementation (Langergraber 2011; Meyer *et al.* 2015). As an example of simple models, the reader is referred to the simulation tools developed to support the design of combined sewer overflow wetlands (Meyer & Dittmer 2015; Palfy *et al.* 2018).

Simplified mechanistic models widely used in chemical engineering have been adopted to model the solute transport as well as removal processes in FWS wetlands. The first models used were the continuously stirred tank reactor (CSTR) and the plug-flow (PF) model (Werner & Kadlec 2000). In their basic form, these models represent ideal reactors with a single residence time. In wetlands, by their nature, effects such as irregular geometry or the turbulence caused by vegetation or the presence of animals cause dispersion transport. Their hydraulic behaviour is characterised by non-ideal flow with mixing zones, dead zones and different velocities at different depths, which result in different residence times (Werner & Kadlec 1996; Walker 1998). In order to simulate the residence time distribution observed in FWS wetlands, models such as the tanks-in-series (TIS) or the plug-flow with dispersion (PFD) have been widely used (Kadlec 1994, 2000; Persson *et al.* 1999; Werner & Kadlec 2000). Also, combined or hybrid models have been used, as is the case with the series-parallel network of tanks model by Kadlec (1994). The hydraulic models can include kinetic expressions for predicting the average behaviour of the system, such as first-order degradation kinetics or Monod kinetics. A widely accepted degradation model is the modified first-order model (k-C* model) that includes background concentrations (C*) (Kadlec 2000). In the k-C* model a compound is never completely removed due to various factors such as uncontrolled inputs and outputs, or internal wetland processes. This model was adapted to the TIS model with an apparent number of tanks (P) to consider pollutant weathering, known as P-k-C* or relaxed TIS, and is currently used for sizing FWS wetlands (Kadlec & Wallace 2009). This model, however, assumes constant detention times during an event and does not account for water and mass losses by evaporation and infiltration. The simple degradation models do not differentiate between microorganisms-degradation and plant-absorption. Some reaction rates are affected by temperature, which can be accounted for in models by adding a correction factor, e.g. Arrhenius equation, as nitrogen compounds.

Several studies have tested simple models to investigate their hydraulic behaviour using performance distributions generated by Monte-Carlo simulations (Werner & Kadlec 2000) or using tracer experiments of existing wetlands (Holland *et al.* 2005; Wong *et al.* 2006; Bodin *et al.* 2012). Many studies performed pilot and large-scale experiments to evaluate the nitrogen removal performance and calculate removal rate coefficients in FWS wetlands (Sun *et al.* 2005; Tuncsiper *et al.* 2006). However, no studies have been found that have compared the models over the entire event using real long-term datasets from large-scale FWS wetlands to identify the limitations of the combined models.

This study aimed to evaluate different FWS wetland models for total nitrogen removal by a full-scale FWS wetland. Various simple models have been applied and evaluated for their ability to predict steady- and unsteady-state conditions.

MATERIALS AND METHODS

Experimental data and boundary conditions

The data used have been collected from a full-scale FWS wetland located at an experimental farm of the Canale Emilio Romagnolo Land Reclamation Consortium in the village of Budrio, Emilia-Romagna region, Italy. It has been in operation and monitored since 2001. According to the Köppen climate classification (Kottek *et al.* 2006), the climate is sub-humid, with an annual average precipitation of 771 mm and an annual average air temperature of 13.7 °C. The wetland has a surface area of 0.4 ha and was fed by a water pump from a channel that collects agricultural runoff from a farm of approximately 12.5 ha in area. The vegetation cover is mostly composed of *Phragmites australis*, *Typha latifolia* and *Carex spp.* The length to width (L:W) aspect ratio was 52, the system being 470 m long and 9 m wide on average. As it can be seen in Figure 1, the FWS wetland has four meanders and the width is not constant. The average depth is 0.4 m, and its total volume is about 1,477 m^3. The bed is permeable, with an infiltration rate ranging between 6.72 and 7.92 mm/d. (Lavrnic *et al.* 2020a).

A number of the inflow events have been recorded by Lavrnic *et al.* (2020b) between 2018 and 2019. Two of these produced outflows and were considered for this study. These are: Event 1, from 1st Feb 2018 to 29th Mar 2018 (56 days), and Event 2, between 17th Nov 2019 and 24th Dec 2019 (37 days). Daily monitoring of precipitation, air temperature, water level, inflow and outflow volume was done during both events. Water samples were taken to measure water quality parameters using two automatic samplers, one at the inlet and one at the outlet. Samples have been collected flow- and time-proportional (Lavrnic *et al.* 2020b). From February to April 2018, additional flow and water level measurements were carried out on an hourly frequency. Additionally, a tracer test was conducted in the wetland using NaCl as a non-reactive tracer to obtain more information on the wetland hydraulics. For more details on the tracer experiment, the reader is referred to Lavrnic *et al.* (2020a).

Water balance

The hydrological behaviour of the FWS wetlands was evaluated using a water balance approach during and shortly before and after Event 1 and Event 2. The water balance was computed as follows:

$$\frac{dV_{(i)}}{dt_{(i)}} = Q_{in(i)} - Q_{out(i)} - I_{(i)} + A\,(P_{(i)}) \tag{1}$$

Figure 1 | Aerial photo of the FWS wetland located in the experimental farm of the Canale Emilio Romagnolo Land Reclamation Consortium, in the village of Budrio, Italy.

where i is the time step, V is the wetland volume over time t, Q_{in}, Q_{out} and I are the inlet flow, outlet flow and infiltration rate in m³/h, respectively, and P the precipitation in m/h on the wetland area A in m² (subsequently called 'direct precipitation', P_{direct}).

A variant of the unsteady-state conditions with simulated data was added to avoid overflow in the model and reduce inconsistencies in the measured values. This also provides an overview of the hydrology depending on the size of the FWS to be constructed. To this end, the water level (h) and the outflow (Q_{out}) in the wetland were calculated for each time step from the water balance according to the following set of equations:

$$V_{(i)} = V_{(i-1)} - I_{(i-1)} + P_{direct(i-1)} + Q_{in(i-1)} - Q_{out(i-1)} \tag{2}$$

$$Q_{out(i)} = \begin{cases} 0, & (V_{(i-1)} \leq V_{max}) \\ Q_{in(i-1)}, & (V_{(i-1)} > V_{max}) \end{cases} \tag{3}$$

$$h_{(i)} = \frac{V_{(i)}}{A * \varepsilon} \tag{4}$$

where P_{direct} is the direct precipitation in m³/h, V_{max} is the maximum volume in m³, calculated as A multiplied by the water level at which outflow occurs (i.e. 0.40 m), and ε, the fraction accessible for water flow, i.e. area not occupied by vegetation, which was 0.9, as estimated by Lavrnic *et al.* (2020a).

Both periods were considered to have negligible evapotranspiration because the events are between November and March, months with very low evapotranspiration. An extension of the simulation of Event 1 was carried out for the month of April in which crop evapotranspiration was considered, calculated from the reference evapotranspiration and the crop coefficient for *Phragmites australis* (Cav.), the most widespread crop in the wetland, assuming standard conditions (Allen *et al.* 1998).

Residence time distribution

The tracer test data were used to determine experimentally the residence time distribution (RTD), i.e. the probability distribution of the time a compound remains in the system (Levenspiel 1999). Its function was used to calculate the average hydraulic retention time and adjust the RTD to different hydraulic models in order to find the best-fit parameter values. Finally, the RTD was used to model solute transport in the PFD model with varying input concentrations and volumes.

Simulation study/model combinations

The hydraulic models were set up for steady- and unsteady-state conditions. For each hydraulic model three degradation models have been applied to simulate the removal of total nitrogen (TN) in the FWS wetland using the experimental data of Events 1 and 2, except for Monod kinetics which was only applied to the CSTR and TIS models. The performance of the models was assessed with root mean square error (RMSE) comparing simulated and observed output concentrations, as it has been proven to be more suitable than other goodness-of-fit measures for simulations comparing concrete data points with simulation results (Ahnert *et al.* 2007).

The following hydraulic models were set up to model different hydraulic behaviours that defined the transport of water and solutes in the wetland:

- **CSTR:** The CSTR model assumes instant and complete mixing within, where the concentration at the outlet was the same as in the wetland (Levenspiel 1999). The transport model consisted of a mass balance performed that considered all inputs and outputs of the system.
- **TIS:** The TIS model consists of several CSTRs in series, emulating dispersion and creating a gamma distribution of residence times (Fogler 2016). HRT and N determine the residence time distribution, where N represents the mixing degree. The estimation of N was done in two ways: (1) by minimising RMSE of the RTD curve with the tracer test, and (2) by dividing the square of the HRT by the variance of the RTD curve as described in Bodin *et al.* (2012).
- **PFD:** The plug-flow model was extended to account for the effect of dispersion using the Peclet number (Pe), a dimensionless term that describes the rate of transport by convection in relation to the rate of transport by diffusion or dispersion (Kadlec 1994; Levenspiel 1999). The effect of dispersion introduced different residence times of the compound in the system. Pe was determined by minimising the RMSE of the RTD curve with the tracer test data.

Three variants were used to compare different flow conditions:

- **Steady-state conditions (#1):** For this variant, the outflow was equal to inflow, whereas infiltration and precipitation on the wetland area were neglected. There was no change in volume. For each event, the nominal HRT was calculated by dividing the average volume and flow rates over the whole event.
- **Unsteady-state conditions with measured data (#2):** This variant used measured water level and outflow data for each time step and included infiltration and direct precipitation on the wetland area. The unsteady state conditions of the PFD model were modelled using the RTD approach, where each time step was treated as a pulse emission (Holland *et al.* 2005).
- **Unsteady-state conditions with simulated data (#3):** A water balance was performed in order to simulate the water level and the outflow in the wetland as explained in the *Water balance* section. The RTD approach was used in the PFD model as in #2.

The degradation models used were:

- **First-order kinetics:** This model is used for removal processes in which the reaction depends linearly on the pollutant concentration (C), where the reaction rate coefficient k determines the velocity of the reaction. C was calculated by dividing the mass (g) by the volume (m^3) of the previous time step to account for volume variation in simulations under unsteady conditions (#2 & #3). We used the median degradation rate at 20 °C (k_{20}) value for TN degradation given by Kadlec & Wallace (2009), i.e. 21.5 m/yr.
- **First-order kinetics with non-zero background concentrations:** A background concentration (C^*) was used to set a limit to the degradation reaction. This approach considers processes that prevent complete removal, such as the transformation of some compounds into others, desorption or washing out of particles from the bottom (Carleton & Montas 2010). We compared C^* of 1.5 mg/L as a reference value (Kadlec & Wallace 2009), and 3.7 mg/L, the lowest measured value of TN at the outlet of Event 1. An additional simulation with a background concentration of 6.3 mg/L was done for Event 2.
- **Monod kinetics:** The Monod equation assumes non-linear growth limited, for example, by a finite substrate or the presence of inhibitory substances. It was used only for CSTR and TIS models. The saturation coefficient (k_s) was set to 0.5 g/m^3 as in the model developed by Langergraber & Simunek (2005), and the maximum reaction rate ($k_{o,\,max}$) they used for denitrification ($k_{o,\,max} = 1\ d^{-1}$) was used to calculate the areal rate constant of total nitrogen degradation as indicated by Kadlec & Wallace (2009).

The effect of temperature was considered in the first-order kinetics with zero and non-zero background concentration using the Arrhenius equation with a temperature correction factor (θ) of 1.056 (Kadlec & Wallace 2009).

RESULTS AND DISCUSSION

Hydrological behaviour

A water balance approach was used in order to evaluate Events 1 and 2. Figure 2 shows the measured data during both events whereby the white areas represent the period considered for the evaluation. Rainfall before Event 1 raised the water level to 10 cm height (Figure 2(a)). Subsequently, the recorded water level in Event 1 rose to 50 cm for 5 days, although the theoretical maximum water level in the wetland was 40 cm. However, no outflow was observed until about 10 days after the inflow began. This was due to changes in the operation of the FWS wetland during this period, which resulted in changes in the outflow and water level. Unexpected outcomes are common in long-term data sets of real systems, as it is difficult to have absolute control, compared to systems under controlled conditions in the laboratory or at a small scale in the field. This led us to use the water balance to simulate the water level and the outflow as another model variant (#3: unsteady-state conditions with simulated data). Before Event 2, the FWS wetland was completely dry and took some time to respond to the inflow as expected (Figure 2(b)).

The outflow/inflow ratio, i.e. the proportion of input water leaving the FWS wetland through the outlet, for Event 1 was 77%, and the mass retention rate of TN was about 36%. During Event 2 the outflow/inflow ratio was only 36%, while the mass retention rate of TN was 74%. This can be explained by the fact that in Event 2 there were days with no outflow between water inflows, increasing the hydraulic retention time in the wetland and therefore TN removal. In their study, Lavrnic *et al.* (2018) reported an irregular water flow in the wetland with yearly outflow/inflow ratios between 9 and 65%. They attributed the water losses to high evapotranspiration during the summer months and infiltration through the bottom and the wetland walls. A considerable amount of water could be retained in the soil matrix. This is particularly important in large wetlands, such as the one used for this study. Being a nature-like system, there may also be present preferential flows formed by the root

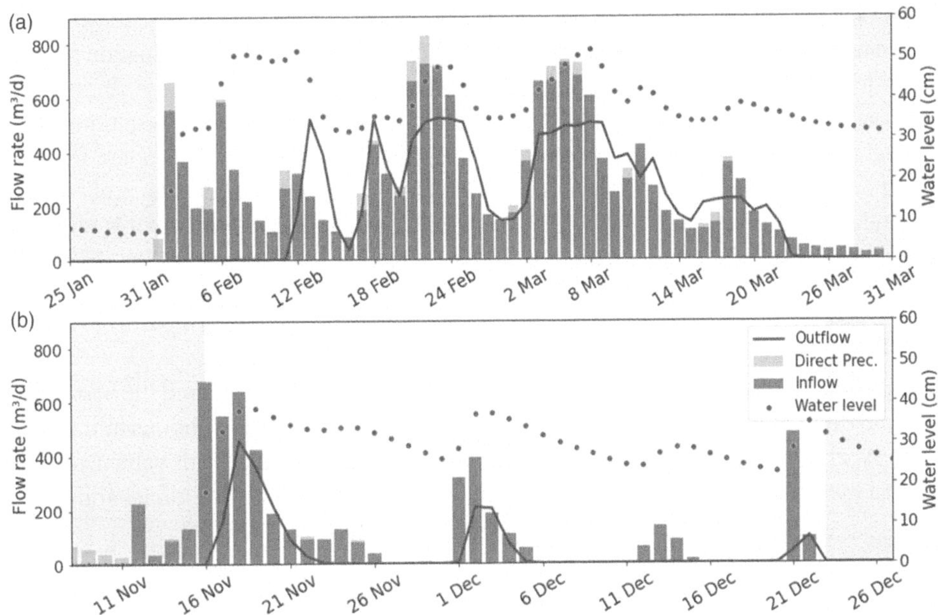

Figure 2 | Overview of measured inflows (inflow and direct precipitation) and outflows (m³/d) and water level (cm) in the wetland during Event 1 (a) and Event 2 (b). The white area represents the period considered in the models (Event 1: 01st Feb 2018–29th Mar 2018; Event 2: 17th Nov 2019–24th Dec 2019).

system, by the presence of macroinvertebrates in the soil or by cracks formed during dry periods. However, soil storage capacity and preferential flows were neglected in our study to keep the model simple.

The measured and simulated values of the water level and outflow during and after Event 1 are plotted in Figure 3. The measured water level has more fluctuations than the simulated water level, which remains below the maximum wetland level, i.e. 40 cm. From the 24th of March onwards, no further inflows or rainfall were measured, and the steady decline in water level was due to infiltration and evapotranspiration in April, when the growing season of *Phragmites australis* began. The decrease was very similar in measured and simulated water levels, and when evapotranspiration was considered, the simulated water level reached the measured values. It is therefore important to take evapotranspiration into account in simulations during warm periods of year.

As shown in Figure 3(b), the simulated outflow started earlier, correlating with the rise in water level. Additionally, the peaks are higher than those of the measured outflow. This can be explained by the fact, that soil storage and other infiltration

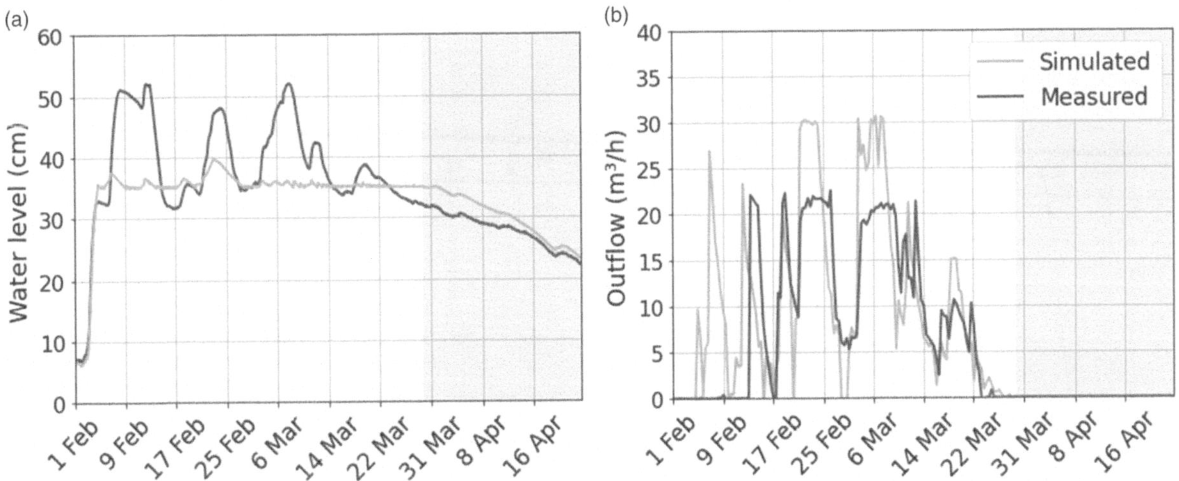

Figure 3 | Measured and simulated water level (a) (cm) and outflow (b) (m³/h) in the wetland during and after Event 1. The white area represents the period included in the models: 01st Feb 2018–29th Mar 2018.

through preferential flows have not been accounted for, the recovery rate using the simulated data would be 100%. Since the simulated outflow of Event 1 started earlier and ended later than the measured values, the evaluation period for this event was extended by a few days for a model comparison covering the whole event.

Figure 4 shows the measured and simulated water level and outflow during Event 2. At the beginning of the event, there was a rapid rise in water level followed by a drop of more than 10 cm in 13 days in the measured values. In contrast, the decline is not as pronounced in the simulated water level. This indicates that in the water balance some processes have not been considered or have been underestimated, such as water storage in the soil or the infiltration rate. It is unlikely that the decrease was due to evapotranspiration since the event occurred between November and December. The simulated outflow was also higher than the measured outflow and lasted longer, similar to Event 1.

Residence time distribution in the hydraulic models

The use of RTD was threefold; (1) to calculate the HRT from the tracer experiment, (2) to find the best-fit parameter values for the transport models, and (3) to model the degradation of pollutants under unsteady conditions in the PFD model. The HRT estimated by the tracer experiment was 6.7 days, as indicated by Lavrnic *et al.* (2020a). This value was used to determine the best-fit parameters in the hydraulic models; the number of tanks-in-series (N) for the TIS model and the Peclet number (Pe) for the PFD model.

TIS model

For an HRT of 6.7 d, the estimated number of tanks was 14, when calculated with the HRT and the variance of the RTD curve. Minimising the RMSE of the RTD curve resulted in an HRT of 8.4 d (RSME of 14.23 mg/L). However, Lavrnic *et al.* (2020a) reported a better fit with 3.78 tanks in series. The mean values reported in the literature for N values in FWS wetlands are 4.1 $\pm$ 0.4 (Kadlec & Wallace 2009). In their study, Persson *et al.* (1999) reported N values between 1.3 and 10, depending on the wetland configuration. In our study, we included 4, 8 and 14 tanks in series to evaluate the model performance with different N in both events.

Figure 5 shows the RTD curves created for different N and the tracer concentrations measured during the experiment. The result for $N = 14$ was more similar to the measured tracer data, although the HRT does not fit the peak of the curve. This tends to occur in wetland systems, where there may be flows with different retention times. In their study, Lavrnic *et al.* (2020a) reported a hydraulic efficiency (e_v) of 0.79 and an effective volume ratio of 0.71, indicating that approximately one-quarter of the wetland did not participate in the active flow.

Plug-flow with dispersion model

The calibration of the PFD model resulted in a Pe of 25 (D/uL = 0.04) for an HRT of 6.7 d, with a RMSE of 10.99 mg/L, the minimum among all hydraulic models applied. The RTD curve fits best at the peak and at the end of the tracer concentration

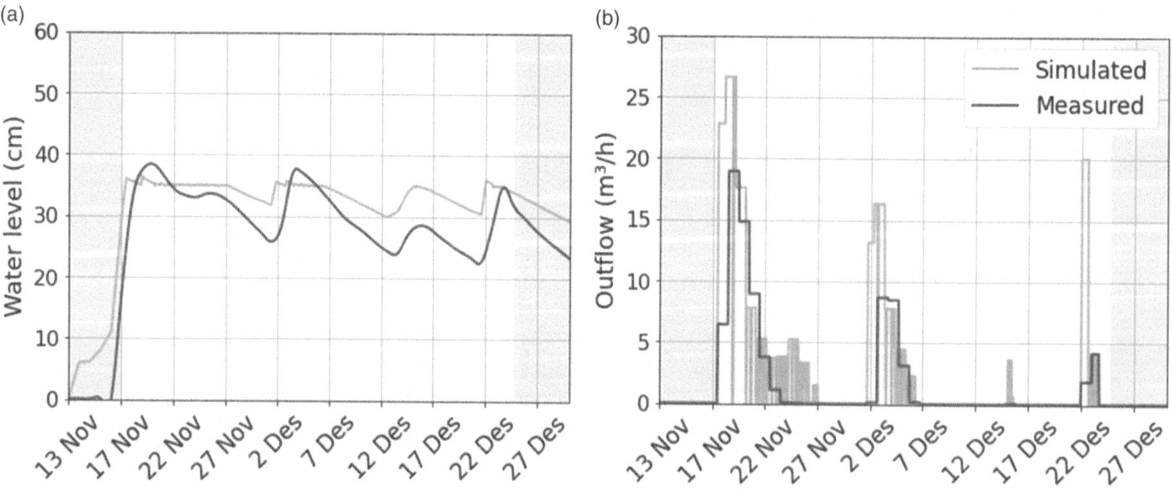

Figure 4 | Measured and simulated water level (a) (cm) and outflow (b) (m³/h) in the wetland in Event 2. The white area represents the period included in the models: 17.11.2019–24.12.2019.

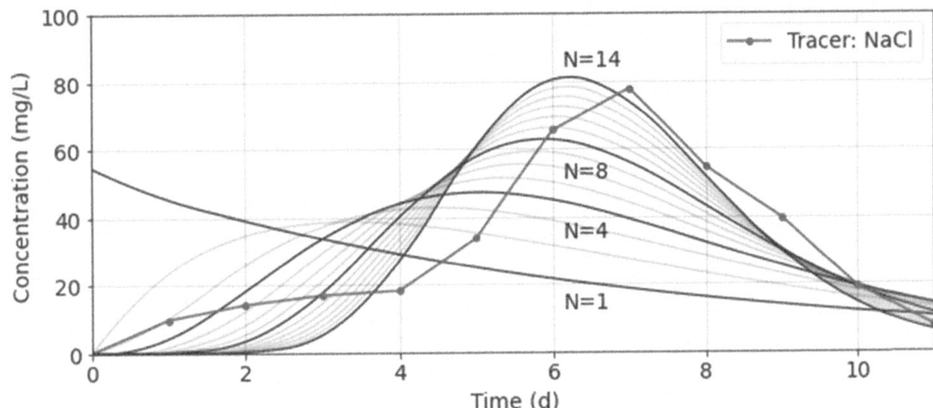

Figure 5 | Residence time distribution curves for the tanks-in-series (TIS) model with different numbers of tanks (*N*) and the measured tracer concentration.

curve but does not seem to fit well at the beginning of the experiment (Figure 6). The results of both models indicate that the wetland had two distinguishable flows: one that arrived earlier to the outlet and could be emulated by a TIS model and another with a longer retention time that could be described as a near-plug flow.

Performance of models for total nitrogen removal

Comparison of hydraulic models

Table 1 gives the RMSE values obtained from the comparison of measured and simulated data for Event 1 and Event 2. All hydraulic models were able to model Total Nitrogen transport with RMSE values between 2.06 and 5.48 mg/L for Event 1 and between 3.17 and 10.62 mg/L for Event 2. The nominal HRT for Event 1 and 2 were 4.7 d and 5.1 d and were used in the PFD model under steady-state conditions (#1). The measured values are represented in the figures as *plateaus* because they were flow- and time-proportional.

When comparing the CSTR and TIS models, CSTR always had higher RMSE values in the first-order degradation kinetics during Event 1. The difference could be attributed to the sharp increase of effluent TN concentration in the CSTR model at the beginning of the event (Figure 7(a)). CSTR models a single tank and assumes instantaneous, complete mixing, and hence the concentration increased immediately at the beginning of the event. This occurred regardless of the degradation model used. By contrast, during Event 2, the modelling of solute transport with a single tank led to better results in most simulations compared to TIS with 4, 8 and 14 tanks-in-series. However, a visual analysis shows that the difference is very small. A better fit was obtained with the TIS models, particularly with $N = 14$ and non-zero background concentrations (Figure 7(b)).

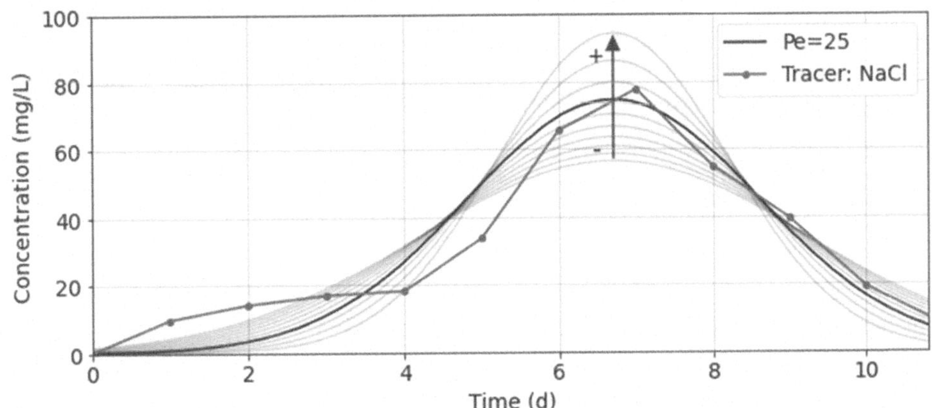

Figure 6 | Residence time distribution curves for the plug-flow with dispersion (PFD) model with a hydraulic retention time of 6.7 d and the measured tracer concentration. The black line indicates the curve with the best fitting Peclet number (*Pe* = 25). The arrow points out the change in the shape of the distribution curve as *Pe* increases.

Table 1 | Root mean squared error (RMSE) values during Event 1 and Event 2 with the combinations of hydraulic and degradation models

				Event 1			Event 2		
				#1	#2	#3	#1	#2	#3
Degradation model	Hydraulic model	C* mg/L	N –	RMSE (mg/L)					
1st order kinetics	CSTR	0	1	2.72	4.95	2.95	4.90	4.13	5.15
	TIS		4	2.36	3.31	2.55	5.19	4.34	5.85
	TIS		8	2.38	2.92	2.59	5.33	4.60	6.19
	TIS		14	2.42	3.28	2.64	5.52	4.82	6.48
	PFD		∞	2.10	5.19	4.78	7.96	10.44	10.62
1st order with non-zero background concentration	CSTR	1.5	1	2.79	5.13	2.96	4.41	3.99	4.68
	TIS		4	2.30	3.51	2.43	4.51	3.67	5.09
	TIS		8	2.28	2.91	2.43	4.60	3.77	5.37
	TIS		14	2.29	3.00	2.46	4.76	3.89	5.64
	PFD		∞	2.06	5.16	4.72	7.14	10.22	10.36
	CSTR	3.7	1	2.99	5.48	3.12	3.88	4.19	4.17
	TIS		4	2.48	4.00	2.48	3.69	3.37	4.14
	TIS		8	2.43	3.22	2.42	3.68	3.21	4.32
	TIS		14	2.40	2.98	2.42	3.78	3.17	4.53
	PFD		∞	2.13	5.21	4.72	6.02	9.90	9.99
Monod kinetics	CSTR		1	2.14	4.11	2.74	5.09	7.44	5.28
	TIS		4	2.17	3.50	2.56	4.94	4.97	5.86
	TIS		8	2.38	3.78	2.69	5.06	5.60	6.21
	TIS		14	2.52	4.46	2.76	5.30	6.26	6.55

C* is the background concentration; N is the number of tanks-in-series.
Hydraulic models: Continuous-stirred tank reactor (CSTR), tanks-in-series (TIS), plug-flow with dispersion (PFD).
Model variants: steady (#1) and unsteady-state conditions using measured (#2) and simulated (#3) water level and outflow data.

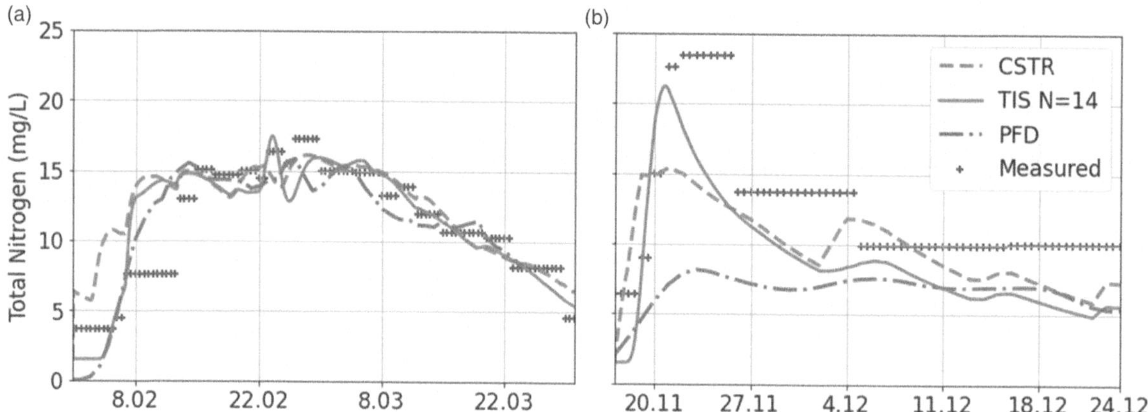

Figure 7 | Measured and simulated total nitrogen effluent concentrations during Event 1 (a) and Event 2 (b) with continuous-stirred tank reactor (CSTR), tanks-in-series (TIS) (N = 14) and plug-flow with dispersion (PFD) models, under steady-state conditions, and first-order degradation kinetics with a background concentration of 1.5 mg/L.

As for the TIS model, a gradual improvement in the simulations was expected to be observed as N increased due to the large length to width ratio of the wetland. However, the number of tanks-in-series in the TIS model did not significantly improve in Event 1. At the beginning of the event, the simulated effluent concentration increased faster than the measured concentration in all simulations. In contrast, the simulated concentrations were close to the measured values during and at the end of the event (Figure 7(a)). Likewise, the differences in RMSE values between 4, 8 or 14 tanks-in-series in Event 1 were small, with standard deviations of 0.02 mg/L, 0.29 mg/L and 0.03 mg/L for variants #1, #2, and #3, respectively. During Event 2, the differences with different N in TIS, and in hydraulic models, in general, are more pronounced. The standard deviations were also slightly larger than in Event 1 (0.09 mg/L, 0.13 mg/L and 0.22 mg/L), with the lowest RMSEs in the models with $N = 4$. However, in both events, it was observed that with a background concentration of 3.7 mg/L, the RMSE value decreased with an increase of N. The FWS wetland used for this study had an elongated shape (L:W = 52), resulting in good hydraulic efficiency ($e_v = 0.79$), so it was expected to perform better for a $N = 14$. In the design of newly constructed or modified FWS wetlands, the hydraulic efficiency can be improved by lengthening the wetland or by adding barriers or other elements within the wetland (Persson *et al.* 1999). Depending on the wetland configuration, a different N could be assumed in the TIS model (Wong *et al.* 2006).

The differences in the PFD model are more noticeable between the events. During Event 1, the PFD model under steady-state conditions (#1) was able to simulate the TN concentration at the wetland outlet with RMSE values of 2.10 ± 0.03 mg/L, the lowest of all combinations and of both events, whereas in Event 2, the PFD model showed the worst performance for all variants and degradation models with RMSEs of 9.18 ± 1.60 mg/L. As shown in Figure 7(b), the simulated output concentration in Event 2 was much lower than the measured concentration and did not even follow sits pattern. The difference between events lies in the outflow rate and intermittency that were a consequence of different precipitation. As seen previously, the outflow in Event 1 was continuous, while in Event 2 it was intermittent. It is precisely the stochastic inputs to the FWS wetlands that treat agricultural runoff that pose a constraint to the PFD model. Irregular inflows lead to different flow velocities affecting dispersive transport, which differ greatly from event to event. This was noticeable when comparing the RTD curve of the tracer experiment with the events. In the tracer test, the lowest RMSE (10.99 mg/L) was obtained for a HRT of 6.7 d with a Pe of 25. However, in Event 1, a better fit could be achieved for a Pe of 1.02 (HRT = 4.7 d; $C^* = 3.7$ mg/L), resulting in RMSE of 1.27 mg/L. Conversely, during Event 2 the Pe resulting from the tracer test was too small, and the RMSE decreased to 3.92 mg/L for a Peclet number of 1335 under steady-state conditions. This confirms that the use of PFD would not be appropriate to model intermittently fed FWS wetlands, even when their configuration resembles a channel.

Steady and unsteady state conditions

The differences between the models depended largely on the state conditions. Under unsteady-state conditions, as opposed to steady-state, fluctuations increased and were not able to simulate the outflow concentration. In Event 1, all models showed better performance under steady-state conditions (#1), followed by unsteady state conditions with simulated water level and outflow (#3), and unsteady conditions with measured data (#2) (Figure 8(a)). In the TIS model, #2 variant had a sharp increase at the beginning, which could not resemble the concentration observed in the wetland. The great peak of TN corresponded to a drastic decrease in the measured water level in the wetland that lead to the concentration of this element in available water volume. Although we used an areal rate constant k_A in the degradation models, the concentration in #2 and #3 was calculated as the mass divided by the volume, which was determined by the water level. Thus, the decrease in water level reduced the degradation of TN and increased the concentration at the outlet. Compared to #1, variant #3 had a slightly lower outflow concentration, which can be attributed to the mass infiltrating into the ground.

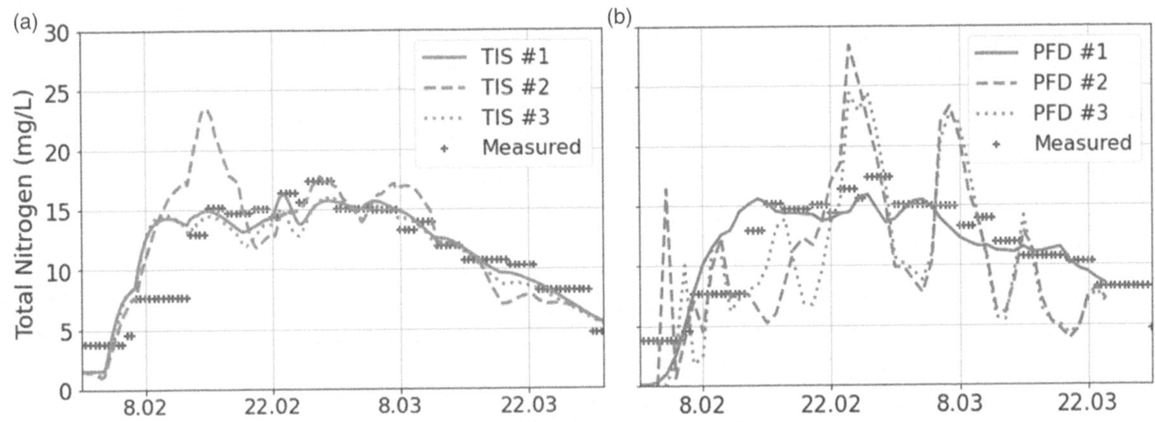

Figure 8 | Measured and simulated total nitrogen effluent concentrations in Event 1 with tanks-in-series (TIS) ($N = 4$) (a) and plug-flow with dispersion (PFD) (b) models under three variants of state conditions: steady-state (#1), unsteady-state with measured (#2) and simulated water level and outflow (#3). The degradation kinetics is first-order with a background concentration of 1.5 mg/L.

In TIS model variant #3, the simulated wetland water level values smoothed the curve and minimised the changes in water level. Nevertheless, this improvement was not as pronounced in variant #3 with the PFD model (Figure 8(b)). In fact, all PFD model simulations under unsteady-state conditions (#2 & #3) resulted in large fluctuations in the effluent concentration that correlated with the input volumes. This can be explained by the influence of the inflow rate on the HRT calculated at each time step, where high inflow rates resulted in lower HRT, entailing a reduction in TN degradation. This supports the findings of Kadlec (2000), who concluded that the calibrated k values tend to increase in the PFD model as HRT decreases.

In contrast to Event 1, the TIS model under unsteady-state conditions (#2) in Event 2 resulted in lower RMSE values for the TIS model (Table 1). However, Figure 9(a) shows the numerical instability in variant #2. This may be due to the high hydraulic variability of this event, with three main discharges and zero outflow in between. Reducing the time step would likely lead to a more stable numerical solution.

Concerning the PFD model in Event 2, none of the variants could model the TN concentration at the wetland outlet, as seen in Figure 9(b). The steady-state variant #1 could simulate output concentrations of the same order of magnitude by increasing Pe. Increasing Pe in #2 and #3 increased the oscillation of the curve, further worsening the result (data not shown).

Comparison of degradation models

All degradation models were able to simulate the increase in concentration, the peak platform and the decrease in Event 1. However, in Event 2, total nitrogen degradation was overestimated in all simulations. The first-order rate constant k was shown to be valid for different events when the effect of temperature was corrected with the Arrhenius equation, even if only air and not water temperature data were available.

Among the degradation models, the Monod-type kinetic gave the best results for Event 1, with the lowest RMSEs in CSTR and TIS with $N = 4$ under steady conditions (Table 1). This was due to a less pronounced increase of the output concentration in the first part of the event compared to the first-order kinetics, as observed in Figure 10. The first-order degradation model, with and without background concentration, simulated a rapid increase in TN concentration at the wetland outlet at the beginning of the event. During the event, an increase of 1 mg/L in the background concentration translated into an increase of approximately 3% in the output concentration. However, the most noticeable difference in the simulated TN concentration was observed rather after Event 1 (since the 29th of March). In Figure 10, the simulations were extended by one month (shaded area). It shows that the combined effect of the first-order degradation kinetics with a background concentration of 3.7 mg/L reproduced the behaviour of the FWS wetland during the latter part of the event, while Monod-type kinetics dropped to zero a few days after the event. In this case, we took the minimum concentration observed in the data set as the background concentration (3.7 mg/L), but this data is not always available. Even so, adding a C^* of 1.5 mg/L resulted in lower RMSE values than the first-order kinetics with zero background concentration.

In Event 2, the first-order degradation model gave better results regardless of the hydraulic model and state conditions, especially when the background concentration was 3.7 mg/L. Interestingly, an additional simulation for Event 2 with a

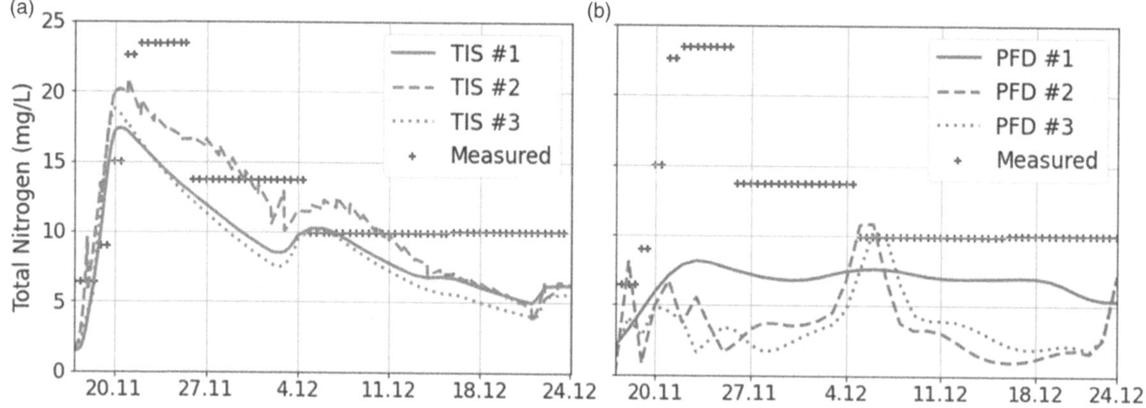

Figure 9 | Measured and simulated total nitrogen effluent concentrations in Event 2 with tanks-in-series (TIS) (a) ($N = 4$) and plug-flow with dispersion (PFD) (b) models under three variants of state conditions: steady-state (#1), unsteady-state with measured (#2) and simulated water level and outflow (#3). The degradation kinetics is first-order with a background concentration of 1.5 mg/L.

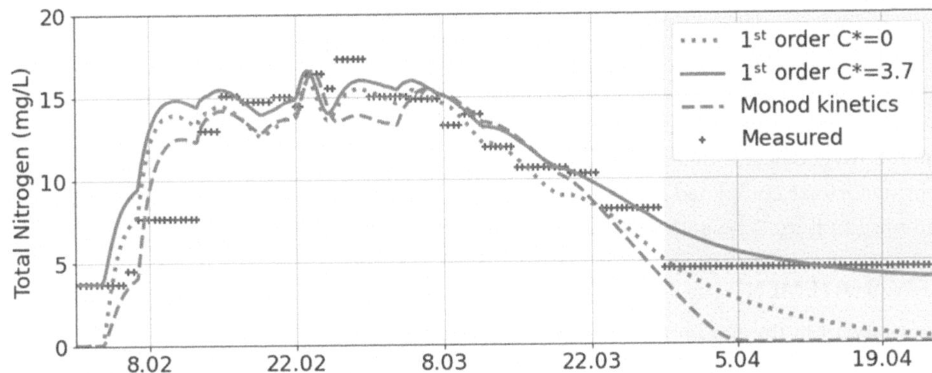

Figure 10 | Measured and simulated total nitrogen effluent concentrations in Event 1 with first-order degradation kinetics with zero and non-zero background concentrations (C^*) and Monod kinetics. The hydraulic model is a tanks-in-series (TIS) model under steady-state conditions ($N = 4$). The shaded area indicates the extended period.

higher background concentration ($C^* = 6.3$ mg/L, the lowest concentration during the event) in the TIS model was shown to improve the performance of the model by 20%. The best fit in the TIS model was observed when $N = 14$ (RMSE $= 3.08$ mg/L). The above shows that the optimal background concentration was different for each event, which could be considered a limitation of the model. Nevertheless, it can be exploited in the model by setting performance limits at different loading conditions using different background concentrations.

In his study, Kadlec (2000) reported that C^* was very close to zero for compounds such as ammonium and nitrate but not for organic nitrogen. In our study, total nitrogen was modelled, which encompasses both nitrogen compounds, so the high C^* value could be due to the proportion of organic nitrogen. Indeed, in the water samples analysed from the inlet, nitrate accounted for 50–90% of the total nitrogen. The highest percentage (90%) was observed when TN concentration was higher, and the lowest (50%) at the end, when TN concentration was lower and the influence of C^* was greater (Figure 10).

CONCLUSIONS

Based on the results of evaluating the model combinations for Events 1 and 2, the following conclusions can be drawn:

- The TIS model proved to be more robust than the CSTR and the plug-flow with dispersion models. Thus, the TIS model is recommended for wetland design, even if the wetland has a large aspect ratio (e.g., as in this study, 52:1).
- The number of tanks, N, will depend on the configuration of the FWS wetland to be designed. The TIS model with $N = 14$ was able to simulate both events well in the FWS wetland modelled in this study, which was quite long. For rather rectangular or uneven FWS wetlands, N close to 4 can be used.
- Although the system has event-dependent inputs (i.e. unsteady-state conditions), the models under steady-state conditions were able to better simulate the total nitrogen outlet concentration of the FWS wetland. The use of simulated outflow and water level data also gave very good results with the TIS model, with an hourly time discretisation.
- Depending on the environmental conditions and purpose, it may be appropriate to use steady or unsteady-state conditions for design purposes. For example, if we expect large evapotranspiration losses or a high infiltration rate, it is desirable to use unsteady-state conditions with simulated data (as in #3 of this paper).
- The first-order kinetics with non-zero background concentration showed the best performance among the degradation models.
- Optimal background concentration varied between events for the same FWS wetland. For the design model, an appropriate background concentration should be included, or a range with lower and upper thresholds can be defined to describe the variability of treatment performance of the FWS wetland.
- The background concentration thresholds should be further explored, as both the current and previous environmental conditions of the FWS wetland (e.g. long periods of drought previous to the event) may influence the background concentration. A background concentration of 1.5 mg/L can be used as a lower limit.

ACKNOWLEDGEMENTS

This work was carried out within the framework of the WATERAGRI (WATER retention and nutrient recycling in soils and streams for improved AGRIcultural production) project, which received funding from the European Union's Horizon 2020 research and innovation programme under grant agreement No 858375. The work of Alba Canet-Martí was supported by the Doctoral School 'Human River Systems in the 21st Century (HR21)' of the University of Natural Resources and Life Sciences, Vienna (BOKU). Sabrina Grüner completed her master's thesis at BOKU and the University of Hohenheim within the European Master Programme EnvEuro. The authors are grateful for the support.

DATA AVAILABILITY STATEMENT

All relevant data are included in the paper or its Supplementary Information.

CONFLICT OF INTEREST STATEMENT

The authors declare there is no conflict.

REFERENCES

Aboukila, A. F. & Deng, Z. Q. 2018 Two variable residence time-based models for removal of total phosphorus and ammonium in free-water surface wetlands. *Ecological Engineering* **111**, 51–59.

Ahnert, M., Blumensaat, F., Langergraber, G., Alex, J., Woerner, D., Frehmann, T., Halft, N., Hobus, I., Plattes, M., Spering, V. & Winkler, S. 2007 Goodness-of-fit measures for numerical modelling in urban water management – a review to support practical applications. In: *Proceedings of the 10th IWA Specialised Conference on 'Design, Operation and Economics of Large Wastewater Treatment Plants'*. IWA (ed.), Vienna, Austria, pp. 69–72.

Allen, R. G., Pereira, L. S., Raes, D. & Smith, M. 1998 *Crop Evapotranspiration: Guidelines for Computing Crop Water Requirements*. Food and Agriculture Organization of the United Nations, Rome.

Aragones, D. G., Sanchez-Ramos, D. & Calvo, G. F. 2020 SURFWET: a biokinetic model for surface flow constructed wetlands. *Science of the Total Environment* **723**, 137650.

Bodin, H., Mietto, A., Ehde, P. M., Persson, J. & Weisner, S. E. B. 2012 Tracer behaviour and analysis of hydraulics in experimental free water surface wetlands. *Ecological Engineering* **49**, 201–211.

Brinson, M. M., Lugo, A. E. & Brown, S. 1981 Primary productivity, decomposition and consumer activity in fresh-water wetlands. *Annual Review of Ecology and Systematics* **12**, 123–161.

Carleton, J. N. & Montas, H. J. 2010 An analysis of performance models for free water surface wetlands. *Water Research* **44** (12), 3595–3606.

Dotro, G., Langergraber, G., Molle, P., Nivala, J., Puigagut, J., Stein, O. & von Sperling, M. 2017 *Treatment Wetlands: Biological Wastewater Treatment Series*, Vol. 7. IWA Publishing, London, UK.

Fogler, H. S. 2016 *Elements of Chemical Reaction Engineering*. Prentice Hall, Kendallville, Indiana, USA.

Gargallo, S., Martin, M., Oliver, N. & Hernandez-Crespo, C. 2017 Biokinetic model for nitrogen removal in free water surface constructed wetlands. *Science of the Total Environment* **587**, 145–156.

Hammer, D. E. & Kadlec, R. H. 1986 A model for wetland surface-water dynamics. *Water Resources Research* **22** (13), 1951–1958.

Holland, J. F., Martin, J. F., Granata, T., Bouchard, V., Quigley, M. & Brown, L. 2005 Analysis and modeling of suspended solids from high-frequency monitoring in a stormwater treatment wetland. *Ecological Engineering* **24** (3), 159–176.

Kadlec, R. H. 1994 Detention and mixing in free-water wetlands. *Ecological Engineering* **3** (4), 345–380.

Kadlec, R. H. 2000 The inadequacy of first-order treatment wetland models. *Ecological Engineering* **15** (1–2), 105–119.

Kadlec, R. H. & Wallace, S. 2009 *Treatment Wetlands*. Taylor & Francis Group, LLC, Boca Raton, FL, USA.

Kottek, M., Grieser, J., Beck, C., Rudolf, B. & Rubel, F. 2006 World map of the Koppen-Geiger climate classification updated. *Meteorologische Zeitschrift* **15** (3), 259–263.

Langergraber, G. 2011 Numerical modelling: a tool for better constructed wetland design? *Water Science and Technology* **64** (1), 14–21.

Langergraber, G. & Simunek, J. 2005 Modeling variably saturated water flow and multicomponent reactive transport in constructed wetlands. *Vadose Zone Journal* **4** (4), 924–938.

Langergraber, G., Dotro, G., Nivala, J., Rizzo, A. & Stein, O. R. 2019 *Wetland Technology. Practical Information on the Design and Application of Treatment Wetlands*, 1st edn. IWA Publishing, London, UK.

Lavrnic, S., Braschi, I., Anconelli, S., Blasioli, S., Solimando, D., Mannini, P. & Toscano, A. 2018 Long-term monitoring of a surface flow constructed wetland treating agricultural drainage water in Northern Italy. *Water* **10** (5), 644.

Lavrnic, S., Alagna, V., Iovino, M., Anconelli, S., Solimando, D. & Toscano, A. 2020a Hydrological and hydraulic behaviour of a surface flow constructed wetland treating agricultural drainage water in northern Italy. *Science of the Total Environment* **702**, 134795.

Lavrnic, S., Nan, X., Blasioli, S., Braschi, I., Anconelli, S. & Toscano, A. 2020b Performance of a full scale constructed wetland as ecological practice for agricultural drainage water treatment in Northern Italy. *Ecological Engineering* **154**, 105927.

Levenspiel, O. 1999 *Chemical Reaction Engineering*. John Wiley & Sons, New York.

Mancuso, G., Bencresciuto, G. F., Lavrnic, S. & Toscano, A. 2021 Diffuse water pollution from agriculture: a review of nature-based solutions for nitrogen removal and recovery. *Water* **13** (14), 1893.

Meyer, D. & Dittmer, U. 2015 RSF_sim – a simulation tool to support the design of constructed wetlands for combined sewer overflow treatment. *Ecological Engineering* **80**, 198–204.

Meyer, D., Chazarenc, F., Claveau-Mallet, D., Dittmer, U., Forquet, N., Molle, P., Morvannou, A., Palfy, T., Petitjean, A., Rizzo, A., Campa, R. S., Scholz, M., Soric, A. & Langergraber, G. 2015 Modelling constructed wetlands: scopes and aims – a comparative review. *Ecological Engineering* **80**, 205–213.

Palfy, T. G., Meyer, D., Troesch, S., Gourdon, R., Olivier, L. & Molle, P. 2018 A single-output model for the dynamic design of constructed wetlands treating combined sewer overflow. *Environmental Modelling & Software* **102**, 49–72.

Persson, J., Somes, N. L. G. & Wong, T. H. F. 1999 Hydraulics efficiency of constructed wetlands and ponds. *Water Science and Technology* **40** (3), 291–300.

Robertson, G. P. & Vitousek, P. M. 2009 Nitrogen in agriculture: balancing the cost of an essential resource. *Annual Review of Environment and Resources* **34**, 97–125.

Stone, K. C., Poach, M. E., Hunt, P. G. & Reddy, G. B. 2004 Marsh-pond-marsh constructed wetland design analysis for swine lagoon wastewater treatment. *Ecological Engineering* **23** (2), 127–133.

Sun, G. Z., Zhao, Y. Q. & Allen, S. 2005 Enhanced removal of organic matter and ammoniacal-nitrogen in a column experiment of tidal flow constructed wetland system. *Journal of Biotechnology* **115** (2), 189–197.

Tuncsiper, B., Ayaz, S. C. & Akca, L. 2006 Modelling and evaluation of nitrogen removal performance in subsurface flow and free water surface constructed wetlands. *Water Science and Technology* **53** (12), 111–120.

Walker, D. J. 1998 Modelling residence time in stormwater ponds. *Ecological Engineering* **10** (3), 247–262.

Werner, T. M. & Kadlec, R. H. 1996 Application of residence time distributions to stormwater treatment systems. *Ecological Engineering* **7** (3), 213–234.

Werner, T. M. & Kadlec, R. H. 2000 Stochastic simulation of partially-mixed, event-driven treatment wetlands. *Ecological Engineering* **14** (3), 253–267.

Withers, P. J. A., Neal, C., Jarvie, H. P. & Doody, D. G. 2014 Agriculture and eutrophication: where do we go from here? *Sustainability* **6** (9), 5853–5875.

Wong, T. H. F., Fletcher, T. D., Duncan, H. P. & Jenkins, G. A. 2006 Modelling urban stormwater treatment – a unified approach. *Ecological Engineering* **27** (1), 58–70.

First received 29 March 2022; accepted in revised form 26 May 2022. Available online 1 June 2022

doi: 10.2166/wst.2022.159

The regime of constructed wetlands in greywater treatment

Ushani Uthirakrishnan[a,*], Vineeth Manthapuri[b], Afrah Harafan[b], Padmanaban Velayudhaperumal Chellam[c] and Tamilarasan Karuppiah[d]

[a] Department of Biotechnology, Karpaga Vinayaga College of Engineering and Technology, Chengalpattu, Tamil Nadu 603 308, India
[b] Environmental & Water Resources Engineering, Department of Civil Engineering, Indian Institute of Technology Madras, Chennai, TN 600036, India
[c] Department of Biotechnology, National Institute of Technology, Andhrapradesh, Tadepalligudem, Andhra Pradesh, India
[d] Department of Civil Engineering, Vel Tech Rangarajan Dr. Sagunthala R&D Institute of Science and Technology, Avadi, Tamil Nadu 600 062, India
*Corresponding author. E-mail: dr.ushaniu@gmail.com

ABSTRACT

There is an excellent need for supply-side threats due to the enhanced degradation and reclamation of existing water bodies in the present scenario. This led to the global water crisis. One of the easiest ways to fulfil the growing need for freshwater is the recycling of wastewater. Greywater is a form of wastewater from households, industries, etc., with some less toxic materials. The recycling of this greywater has provoked the development of new and sustainable technologies to meet the growing water demand. Engineered constructed wetlands are considered one of the most economically practical processes to treat greywater due to its minimal footprint. In this case study, we summarize several categories of constructed wetlands, operating conditions, and the effects of biological, physical, and chemical aspects of greywater on their treatment performance. On the other hand, the effluent quality from diverse wetlands is also summarized. Furthermore, it would be better to consider that constructed wetlands' integrated performance with disinfection may improve the effluent quality to desirable standards.

Key words: green roofs, green walls, greywater, hybrid constructed wetlands, wetland

HIGHLIGHTS

- Constructed wetlands' integrated practices with disinfection improve the effluent quality.
- To fulfill the need for freshwater recycling of greywater is necessary.
- Engineered constructed wetlands are economically practical to treat greywater.

GRAPHICAL ABSTRACT

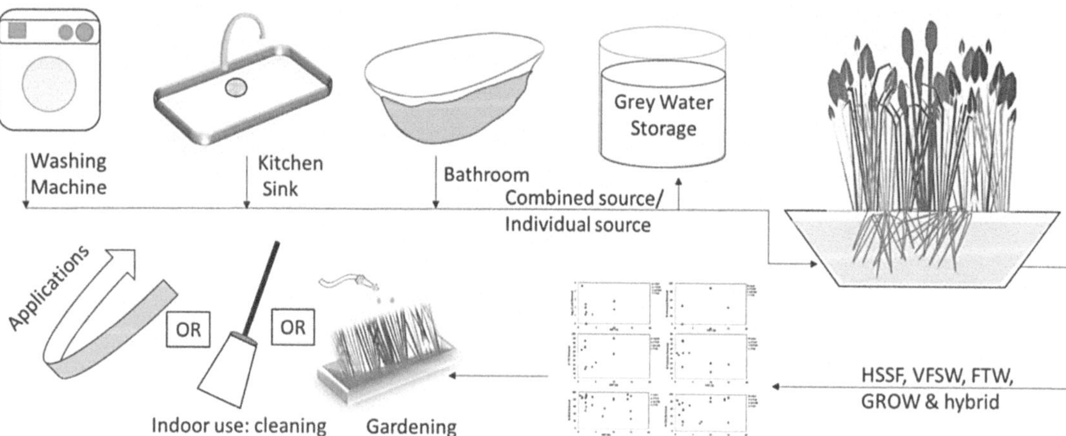

INTRODUCTION

An era of boundless innovations with technical advancement has been achieved by the human race every day. The introduction of machines has evoked new trades, production policies, more mechanized manual production, and reduction in production time. Both the developing and developed nations were pushed to face social and environmental disbalances after the industrialization. Redclift (2009) researched upon the insignificance of economic progression with non-economic factors. Although urbanization and rapid industrialization been the marks of an emerging country, the consequences like water demand and water scarcity are on the rise in a progressive gradient. By 2050 around the globe there will be a 55% upsurge in the water necessity according to the report of the United Nations World Water Development 2015. The well-being of humans, survival of ecology, and economic pursuits are wholly dependent upon the accessibility of water (Kounina *et al.* 2013). Of the total surface of earth about 70% constitutes water where only 3% of it is available in the form of fresh water making it a most valuable resource. According to a report from 1966, throughout global dimension a gigantic decrease of 55% is observed in the availability of freshwater. (UNU-INWEH 2017). Contrastingly, as a consequence of uncontrollable population increase, most conveniently designed units of wastewater treatment are flowing out of their usual abilities. These planned treatment divisions are capable of handling merely 20% of sewage stock while the leftover 80% are untreated and will be discharged into water bodies acceding to the reports of UNU-INWEH (2017). While on the other side of operation at the global scale into the atmosphere the sewage treatment plants (STPs) liberate 1,400 tons of CO_2 (Muga & Mihelcic 2008). For a better human health and sustainable practices, the long-distance transportation of water should be alleviated. The impediment in the usage of typical driven gravity sewer system around the globe is due to the economic and demographic trends. A shift in paradigm practice is essential for effective utilization of resources and a steady sustainable progress. A significance in reducing the water demand around the globe and the burden on conventional wastewater treatment units are achieved by an alternative source like greywater.

Greywater is said to be one among the other types of wastewaters that emerges from non-toilet wastewater in households. Categorized wastewater under low to a medium category of greywater supposed to have some ups and downs in its own chemical, physical, and biological factors (Friedler & Hadari 2006). A standard of around 100–150 L per capita per day (lpcd) of wastewater has been generated by each and every individual. Of the total produced domestic wastewater, greywater accounts up to 65–85% further can be improved up to 90% by fixing vacuum toilets (Hernández-Leal *et al.* 2011; Penn *et al.* 2012; Ghaitidak & Yadav 2013). After recycling greywater, a decrease of 30–10% was achieved in the portable demand for flushing (Karpiscak *et al.* 1990; Eriksson *et al.* 2002; Vuppaladadiyam *et al.* 2018). When compared to centralized treatment plants, a reduction in the energy load from 37.5% to 11.7% and the minimisation of 25.71% emissions of CO_2 is said to be achieved using greywater decentralized treatment units (Matos *et al.* 2014). The reason to make reprocessing of greywater an interesting and emotive issue is their ability to improve water anxiety and its contents on traditional STPs.

Treatment of greywater before its reuse is very necessary to improve its aesthetical appearance and quality as its stances a greater threat to both plant and human lifespan. Microalgae is suited to be one of the supportable choices for greywater treatment through the world. When compared with other technologies the algae reactors were more economic and easier to operate placing it in a commendable choice. The goal of the current work is to check the efficiency of various treatment that was carried out in greywater treatment.

ANALYSIS AND PROPERTIES OF GREYWATER

Greywater in common is recycled water that originates from lavatory, laundry, and kitchen (Friedler & Hadari 2006; Gross *et al.* 2007; Eriksson *et al.* 2009; Saumya *et al.* 2015). Also, the water that flows without any interaction from lavatory leftover is also termed as greywater (Ghaitidak & Yadav 2013). It is also mentioned as sullage and light wastewater (Morel & Diener 2006). The characteristics of transformation of shade into grey even for a shorter storing period were entitled to its name 'greywater'. Load greywater were additional classified into low pollutant greywater (LGW) and high pollutant greywater (HGW) depending on the pollutant (Boyjoo *et al.* 2013). In the domain of wastewater recycling application greywater is considered to be a valuable source in the likes of high volume and low pollutant load (Oteng-Peprah *et al.* 2018).

Naturally, greywater accords up to 60–85% of total household wastewater (Abed *et al.* 2020), and it can also reach a high percentage of 90–100% in houses practicing dry toilets system (Sheng *et al.* 2020). Developed nations of the USA, Asia, and Germany tend to generate the greywater up to 200 lpcd, 72–225 lpcd, and 35–150 lpcd (Morel & Diener 2006; Mandal *et al.* 2011). Espoused eco-village tradition in various parts of Holland, Germany, Sweden, and Norway impede Germany as one of the lowest greywater producers compared to Asia and the USA. Moreover, in some Africa regions, greywater production is noticed to be on a minimal note of only 14 lpcd due to water scarcity a low economy (Al-Hamaiedeh & Bino 2010; Shafiquzzaman *et al.* 2020). However, numerous desalination plants in Oman (an arid region) contribute up to 160 lpcd, almost 82% conversion of domestic wastewater into greywater is observed (Oh *et al.* 2018). Due to advanced technology and proper water supply, developed nations tend to produce more greywater (Figure 1) (Elhegazy & Eid 2020; Eriksson *et al.* 2002).

Quality and quantity of domestic water supply, distribution systems for both drinking water and greywater, lifestyle activities of the menage, and source of greywater are the prolific influencers of the greywater characteristics across the world (Ghaitidak & Yadav 2013; Oteng-Peprah *et al.* 2018). Areas with substantial quantity and quality of water supply tend to dilute the greywater resulting in a low level of pollutants. Biological and chemical degradation in the biofilm formed, leaching pollutants into the water, and also depends on the type of sewer system used to transport water media. Usage of chemicals in diurnal activities tends to vary with the lifestyle of the menage resulting in variation of characteristics of the greywater across the globe.

The general temperature of greywater is examined to vary from 18 to 38 °C. In general, TSS concentration in greywater varies from 23 to 1,140 mg/L. Turbidity values in the greywater usually vary from (20 to 619 NTU), laundry being the predominant source. pH and alkalinity of the household water supply affect the pH of the grey wastewater. Detergents in cleansing powders elevate the pH of greywater. In general, the pH of grey tends to vary from (6 to 8.2). Low values of pH in greywater are due to the occurrence of amino acids. COD and BOD in the greywater usually vary in the range of (25–2,878 mg/L and 14–1,240 mg/L), respectively. Excessive chemicals for cleansing purposes and food waste are the main reason for the high noticeable concentration of COD and BOD in greywater (Table 1).

Compared with domestic wastewater, greywater deficits in total nitrogen concentration, mainly due to absences of urine in the greywater, however proteins in food and meat make up to a nominal concentration of (1.7–36 mg/L) TN in greywater. Total phosphorus in the greywater is primarily originated from laundry sources due to the extravagant usage of phosphorous containing chemicals. However, in some nations, a shallow concentration of total phosphorus is noticed due to prohibiting phosphorus-containing chemicals (Pinto & Maheshwari 2015). In general, greywater manifests TP variation of 0.11–48 mg/L. Direct contact of greywater after using toilets and washing babies' nappies is the prolific reason for the elevated levels of total

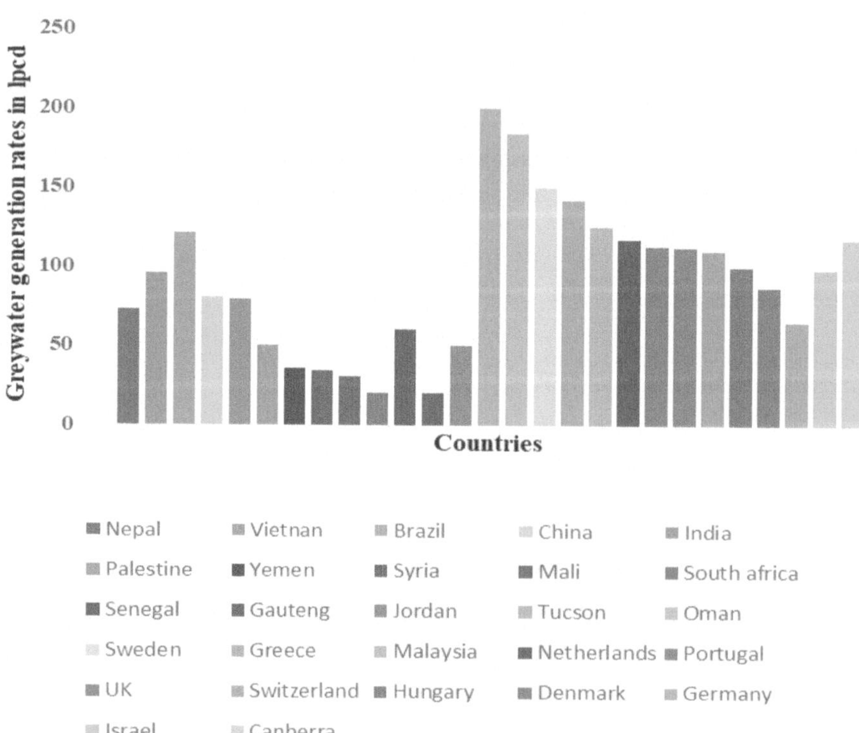

Figure 1 | Greywater generation rates from different countries.

Table 1 | The variation of physico-chemical and biological parameters of greywater

Parameters	Combined source	Bathroom source	Kitchen source	Laundry source
pH	6–8.3	5–8.3	5.58–8	7.3–10
BOD (mg/L)	14–1,240	18–848	100–1,850	38–2,743
COD (mg/L)	25–2,878	77–1,800	26–8,071	58–6,497
TN (mg/L)	1.7–36	0.5–36	1.5–48	0.9–40.3
TP (mg/L)	0.11–48	0.04–11.3	1.4–25	0.062–57
TSS (mg/L)	23–1,140	7–1,160	134–1,300	39.5–3,060
Turbidity (NTU)	20–619	17.1–375	133–298	39–444
Temperature (°C)	18–38	25.8–33	23–30	27–37
TC/100 mL	$56 - 2.67E + 07$	$10 - 1.70E + 06$	$>1.00E + 07$	$200.5 - 2.10E + 06$
FC/100 mL	$0.1 - 1.23E + 05$	$0 - 3.77E + 04$	$>11.60E + 05$	$50.1 - 1.05E + 05$

coliform and faecal coliform in greywater. In general, total TC, FC values vary in the range of $(56 - 2.67E + 07)$ and $(0.1 - 1.23E + 05/100)$ mL, respectively.

CONSTRUCTED WETLANDS (CWS) FOR EFFICIENT TREATMENT OF GREYWATER

Wetlands are naturally occurring distinct eco-system that is temporally or permanently inundated with water (Keddy 2010). These natural bodies serve various purposes, starting from water treatment to home for the diverse nature of animals and birds by maintaining ecological balance (Dorney *et al.* 2018). Contrastingly, constructed wetlands are engineered to mimic the natural phenomena, prolifically treating wastewater in a controlled demeanour. Griffiths & Mitsch (2017) expounded constructed wetlands as a designed, human-made complex of submerged and emergent vegetation along with the saturated substrate, animal life, and water that vivifies wetlands for social benefits and uses. Constructed wetlands are traditionally developed treatment systems with a blend of physical, biological, and chemical mechanisms for sustainable treating of wastewater (Kadlec & Wallace 2008; Zhang *et al.* 2015). Constructed wetland plays a prolific role in water treatment, mainly in areas of tropical and subtropical climates of developing economies (Ali 2018). Compared with conventional centralized sewage treatment plants, wetlands generate a minimal carbon footprint, very economical, and easy to operate (Wang *et al.* 2020). Despite the eco-friendly nature of the CWs, they are widely substituted for secondary and tertiary treatment units in the conventional municipal treatment units and even for greywater treatment. Organic matter, nutrients, and heavy metals are conventionally detached in the wetlands. Even though CWs provide indirect advantages of wildlife habitats, green space, and water purification at nominal operation and maintenance costs, they lack the proper understanding of design aspects.

The conventionally espoused treatment systems for treating either wastewater or greywater are activated sludge process (ASP), sequential batch reactor (SBR), membrane bioreactor (MBR), an up-flow anaerobic sludge blanket (UASB). The significant shortcomings of all this conventional system are the excessive generation of sludge, the involvement of enormous mechanical and electrical components, which eventually ameliorates the high operation and maintains cost and odor and nuisance problem. Even though sludge volume accounts for up to only 1–3% of the total volume of wastewater, the cost involved in the handling and managing sludge may exceed 50% of the total cost. Contrarily, constructed wetlands are cost-efficient treatment technologies with almost similar or less economic inputs. On a global scale, the operation cost incurred during the conventional electrical or mechanical treatment units is 90% more expensive than that of CWs. Unlike conventional treatment units, there is no usage of chemicals, eventually resulting in low sludge volumes. The only sludge generated in CWs is the plant biomass, which can be cleaned more easily and can also be used for biomass production and eventually for biofuel. Moreover, CWs pumps consume an appreciable quantity of energy, which can be alleviated by adjusting the bed slope.

CLASSIFICATION OF CONSTRUCTED WETLANDS

CWs are predominantly classified as four types in the view of water treatment, as shown in Figure 2. The classification (Figure 2) is majorly based on flow directions.

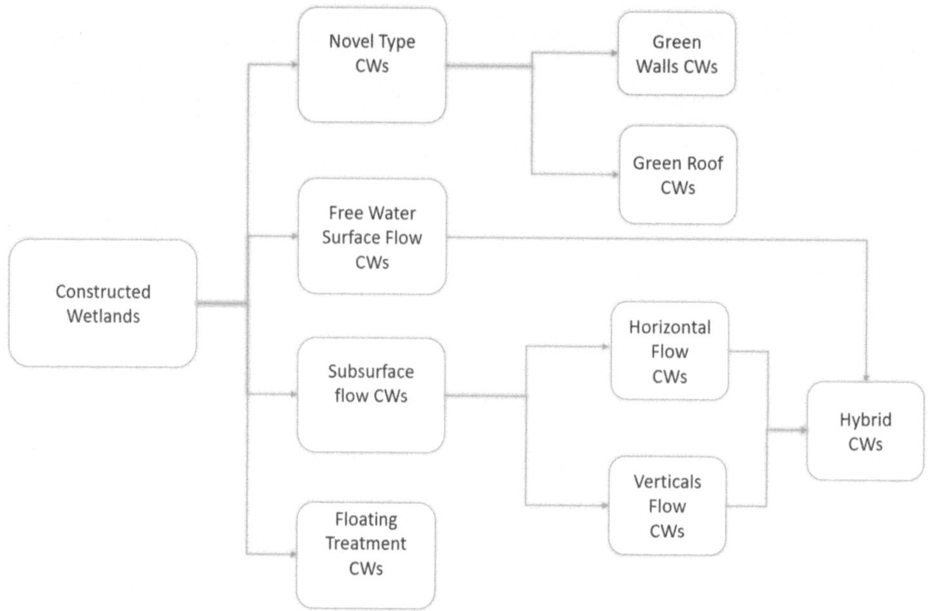

Figure 2 | Classification of constructed wetlands.

Free water surface constructed wetlands (FWS CWs) or surface flow constructed wetlands (SF CWs)

In general, FWS CWs replicates the natural wetlands to a more considerable extent with considerably larger land requirements. FWS CWs are shallow basin engineered wetlands with 10–50 cm of the water surface above the 30–40 cm of substrate media. Usually, soil, the bottom of the soil, is substantially coated with clay material or geomembrane/geotextile to avoid water leaking. Cattails (*Typha* spp.), common reeds (*Phragmites australis*), bulrush (*Scirpus* spp.), and herbs (*Juncus* spp.) are commonly employed plants in FWS CWs for treating the greywater and wastewater (Vymazal 2013; Stefanakis *et al.* 2014). Due to complete water contact with different parts of the plants, a satisfactory removal of pollutants is expected through physical, chemical, and biological methods. The magnificent elimination efficacy of BOD and TSS and appreciable elimination of pathogens and TN is observed for FWS CWS but limits its efficiency in TP removal (Kadlec & Wallace 2008; Vymazal 2013). Mostly FWS CWs are employed in the polishing of treated effluent, agriculture effluents, highway storms, and occasionally for treating hydrocarbons in the water (Kotti *et al.* 2010). Due to practical and uniform utilization of substrate along with breadth, considerable high removal efficiency is observed in comparison with other wetlands.

Horizontal flow constructed wetlands (HFCWs)

Initially, HFCs are mostly practiced in high-income countries like the USA and Germany due to high end of operation cost involved for the substrate. Gravel and rock of different composite and mixture are widely used as a substrate for this type of wetland along with plant species like *Typha* (e.g., latifolia, angustifolia), common reeds (*Phragmites australis*), and *Scirpus* (e.g., lacustris, californicus). The soil's bottom is substantially coated with clay material or geomembrane/geotextile to avoid leaking of water, and a mild slope of 1–3% is maintained. In HFCWs substrate layer varies from 30 to 100 cm in depth with water level maintained well beneath the 5–10 cm from the substrate's free surface, henceforth avoiding free contact of water to the atmosphere. Horizontal flowing water in HFCWs, through the substrate's pores, undergoes physical, chemical, and biological deterioration. HFCWs are very efficient in the removal of organic matter but limits in nutrient removal. Principally, HFCWs have widely opted for industrial effluents, landfill leachate, and contaminated groundwater.

Vertical flow constructed wetlands (VFCWs)

Vertical flow constructed wetlands (VFCWs) are commonly more expensive in comparison with other types of wetlands. Water flows vertically through a substrate comprised of a different layer of sand, gravel, and stones organized in an increasing gradient trend and gets at the bottom through a perforated tube. Due to this perforation, better oxygen transfers across the substrate of VFWCs and results in better effluent than other CWs. VFCWs are usually composed of filter media depth varying from 30 to 180 cm with plant cattails (*Typha latifolia*) and common reeds (*Phragmites australis*) as the typical vegetation. The soil's bottom is substantially coated with clay material or geomembrane/geotextile to avoid leaking of water, and a mild slope of 1–2% is maintained. To avoid the clogging of perforated tubes, intermittent loading operation is performed, and surface areas of VFCWs are small compared to FWSCWs and HFCWs. Due to adequate oxygen transfer across the substrate, fabulous sauce removal trends are noticed in organic matter and ammoniacal nitrogen. VFCWs have widely opted for municipal, domestic and industrial wastewater. Inevitably water level sub surface flow (SSF) is maintained well below the media, which eventually results in evasion of mosquitoes, public contact, and odor for SSF. In general, SSF is very useful and requires fewer footprints in comparison with FWS. However, the substrate's cost in SSF is the major shortcoming for its applicability, primarily in the developing nations (Kadlec & Wallace 2008).

Floating treatment wetlands (FTWs)

FTWs integrate CWs with ponds, a material with a relative density less than one typically used for growing the plants on the water surface, with plant roots expanding into the water. Therefore, the system visually resembling like a floating island and is unaffected by variation of water levels. Unlike the conventional CWs, roots in FTWs are free and have a greater contact area (Walker 2017). *Agrostis alba, Iris ensata, Carex strictavirgatum*, and *Panicum* are few opted plants for FTWs (Spangler *et al.* 2019). The developed bio-film on hanging roots enhances bacterial growth, thereby enhancing the nitrogen sequestration and adsorption of total phosphorus (Lucke *et al.* 2019) and, by retarding the flow FTWs, promote sedimentation and filtration (Shahid *et al.* 2018). Due to their floating characteristics, FTWs are widely used in eutrophicated water bodies.

Green walls and green roof

Huge land acquisition for almost all the constructed wetlands restricts their use to mostly rural and urban fringes. To overcome the above shortcomings in the conventional paradigm practice and increase the adaptability of cost-efficient and eco-friendly treatment processes in highly dense areas, green roof and green walls have been developed (Manso *et al.* 2021). These systems (green walls and green roof) are directly grown on the walls and roof of the buildings directly and they also serve for decorative purpose (Xu *et al.* 2020). Due to their direct growth on walls and roof of the buildings they acquire less space in comparison to other conventionally practiced wetlands (Pradhan *et al.* 2019). In the study of Fowdar *et al.* (2017); treated greywater using green walls with different ornamental plants reported a removal of >90% of BOD >80% of TSS, >99% of phosphorus and >70% of nitrogen over an operating period of one year. In 2020, Xu *et al.* (2020), was successful in treating greywater with green walls, in the study over an operation period of 8 days, green wall system is effective in removing BOD_5 COD, turbidity and anionic surfactants, and turbidity were 97%, 81%, 75%, and 88%, respectively.

   Apart from treating greywater, this system reduces noise pollution, improves air quality, promotes storm water quality, and improves the energy-efficient of the buildings (Berndtsson 2010; Pugh *et al.* 2012). Green walls are usually long vertical growing systems on the interior or exterior walls of the system. The green walls' roots were generally grown over the entire façade (Manso & Castro-Gomes 2015), whereas green roofs are horizontally grown on the top of the roof, and are normally modular systems incorporated with a waterproof membrane, insulation layer, and a vegetation layer rooted in a growing substrate. Depending on the type of vegetation and weight holding capacity of the holding roof, the height of the substrate in the green roof varies from 50 mm to 1 m (Francis & Jensen 2017). *Canna lilies, Carex appressa, Lonicera japonica*, and *Ornamental grapevine* are commonly used plants in green roofs and green walls. The operational phenomena of this novel system are similar to that of conventional CWs but have the additional advantage of a small footprint.

DESIGN OF CONSTRUCTED WETLANDS

Wetland was established in Germany in the 1950s by Kaethe Seidel, who planned the HFCWs, using granular constituents as a grubbing medium. In the 1960s, Reinhold Kickuth investigated granular soil with more clay matter and named the arrangement the 'Root Zone Method.' In the early 1980s, the HFCWs technologies was introduced to Denmark, and by 1987 approximately 100 soil based units were set in process. In the course of the late 1980s, the HFCWs were likewise familiarized to further nations, such as Austria and the UK. In the 1990s, this scheme extended into furthermost Africa, Australia, Asia, European nations, and North America.

One of the significant shortcomings affecting the efficiency of CWs is inadequate knowledge in designing aspects mainly for prodigious CWs (Schulz *et al.* 2019). Construction and design of CWs are usually overlooked and considered simple, resulting in improper and non-sustainable design. As a result of improper design, improper oxygen transfer tends to affect nitrification and denitrification by forming anaerobic bio-film. To date, there are no unanimous sets of guidelines for the design of wetlands; they tend to change from place to place, and depend upon the type of end-use. However, there two basic models developed.

CWs are generally designed based on the removal efficiency of limited parameters like BOD, nutrients removal, TSS, or disinfection performance. The final dimension of the treatment unit is evaluated based on the limiting parameters. Theoretically, the first-order plug model is applied, CWs as attached growth system in steady-state condition with an expected outcome in an exponential manner. Contrastingly, few rules of thumb are developed for designing the small CWs (Vergeles *et al.* 2015). 'Reed' model and 'Kadlec and Knight' model are the most widely adopted models in designing the CWs; both models are predominantly distinguished based on constant rate selection in the plug flow model. In the Reed model, the rate constant is volume and temperature dependant. In the later model rate, the constant is independent of temperature and area-based.

Reed model

The removal efficiency only of nitrate, ammonia, and BOD with the corresponding rate constants as 1.00 day^{-1}, 0.2187 day^{-1}, and 0.678 day^{-1} at 200 °C is calculated using the below Equation (1).

$$ln\left(\frac{C_i}{C_o}\right) = kt \tag{1}$$

On the other hand, (Equation (2)) is used for calculating the removal efficiency of pathogens, TSS, and total phosphorous. Equation (2) is along similar lines to that used for stabilization ponds.

$$\left(\frac{C_i}{C_o}\right) = (1 + kt)^n \tag{2}$$

where

K is the reaction rate constant/day
t is the hydraulic residence time in days
C_i is the influent concentration (mg/L)
C_o is the final expected effluent concentration (mg/L)
n is the number of cells in the series

Kadlec and Knight model

Background concentration of the pollutants along with hydraulic parameters like hydraulic conductivity and hydraulic loading rate was instigated in Kadlec and Knight model (Equation (3)).

$$lnln\left[\frac{C_0 - C^*}{C_i - C^*}\right] = -\frac{k}{q} \tag{3}$$

$$q = \frac{365Q}{A_s} \tag{4}$$

where

C* is background pollutant concentration (mg/L)
C_i is the influent concentration (mg/L)
C_o is the final expected effluent concentration (mg/L)
As is the treatment area of wetland (m^2)

Q is average discharge through the wetland (m^3/day).

q is the hydraulic loading rate (m/year).

CONSTRUCTED WETLANDS IN GREYWATER TREATMENT

The novelty of CWs in treating greywater includes integrating two primary treatment mechanisms of physical and biological degradation of the pollutants. The active substrate in CWs entraps pollutants, and various biotic and abiotic organisms in CWs perform biological degradation (Garcia *et al.* 2010; Jasper *et al.* 2013). From the past few decades of the research, there exists a predilection gap on selecting appropriate greywater treatment from various physical, chemical, and biological processes (Kadlec & Wallace 2008; De Gisi *et al.* 2016). Contrastingly, there is a possibility of occurrence of all the three processes in an eco-friendly manner in CWs. In the present section, an attempt was made to understand the adaptability of CWs for greywater treatment. Systems like VFCWs, HFCWs, hybrid systems, and even novel systems like GROW are also examined in treating greywater.

Horizontal flow constructed wetland (HFCW)

Freezing the water's top surface during the winter is a significant issue while treating the cold countries' wastewater. In such conditions, CWs have opted as secondary treatment. Wang *et al.* (2017) examined the adaptability of HFCWs in a cold region like Sweden. In the study (Wang *et al.* 2017), greywater is treated with HFCWs is preceded by aerobic filters, and good removal efficiency is noticed even during the winter due to long retention time of 6–7 days for HFCWs. The appreciable removal efficiency of 78%, 81.9%, and 50% are observed for TP, BOD7, and TN respectively for CW, which is very high compared to the respective efficiency of the preceding bio-filter. Recycled water in the study (Wang *et al.* 2017) meets the European swimming pool reuse standards.

In the studies of El Hamouri *et al.* (2008), a comparison was performed between low-cost HFCWs supported with multi-layer vertical filter system and extravagant technologies in conventional wastewater treatment units like SBR and MBR with capacities of 600 L/day and 550 L/day, respectively, for treating bathroom greywater of a sports complex. The study (El Hamouri *et al.* 2008) elucidated low-cost HFCWs system planted with Phragmites can reduce the turbidity, BOD, COD, TN, and TP by 93%, 85%, 75%, and 50% respectively, even an appreciable removal efficiency of 90% is noticed for surfactants. The performance of HFCWs along with the filter system is on par SBR and MBR in treating the bathroom greywater for the reuse of toilet flushing for 15 continuous months.

Abdel-Shafy *et al.* (2009) elucidated the efficiency of HFCWs as a secondary treatment unit preceded by UASB in treating the greywater from urban households of Egypt. In comparison with UASB, HFCWs exhibited superior removal efficiency for the integrated system. Noticed removal efficiency for CWs with *Phragmites australis* are 83.5%, 86.4%, 89%, 90%, 95.2%, 69.3%, 56.2% and 99.999% for COD, BOD, TSS, turbidity, sulfides, TKN, TP, and faecal coliform, respectively. Abdel-Shafy *et al.* (2009) also investigated a different fraction of COD, and HFCW effectively decreased suspended COD by 17% is on par with UASB.

Laaffat *et al.* (2015) examined the potential of HFCWs in treating greywater emerging from primary schools in morocco. In the study (Laaffat *et al.* 2015), HFCWs with *Typha latifolia* is preceded by a coarse screen unit for preliminary treatment, and monitoring of the treatment is performed for 100 days. Results of the study manifested a removal efficiency of 92% BOD 5, 85% of COD, 45% of TN, 41% of TP, and 99% of *E. coli* from a respective influent value of 44.2 mg/L of BOD5, 77.2 mg/L of COD, 7.1 mg/L of TN, 0.8 mg/L of TP and 5×10^3 of *E. coli*. Reclaimed greywater is successfully recycled for toilet flushing.

Arden & Ma (2018) reviewed many literature on constructed wetlands for treating greywater with an outcome of average removal efficiency through various analyses. According to the literatures average BOD, TSS, turbidity, TN and TP removal efficiency are 87%, 64%, 47%, 44% and 24% respectively, for the average influent values of 196 mg/L, 52 mg/L, 89 NTU, 7.2 mg/L and 2.7 mg/L. The average removal efficiency of HFCWs is equivalence to VFCWs but less than the hybrid system. Mars *et al.* (2003) conducted various experiments with different HRT of 20 days and 10 days. Various studies were conducted to understand the effect of ponding, HRT, and vegetation in nutrient uptake (Mars *et al.* 2003). Final outcome was compared with SSF and surface flow for a HRT 10 days and all the plant leaves, roots and tubers were digested at end of four months to analyses the phosphorus and nitrogen uptaken by the plants condition after three months of the study. Studies depicted tanks planted with *Triglochinhuegelii* shown better uptake of nutrients, and the subsurface system with vegetation shown appreciable uptake of nutrients in comparison with the surface system with vegetation at the same HRT.

Vertical flow constructed wetlands

Li *et al.* (2004) developed an integrated system by combining VF CWs and photocatalytic oxidation mainly for semi-tropical and tropical climate zones. In the study (Li *et al.* 2004), preliminary treatments grit chambers with solid grease traps are equipped before VF CWs. VF CWs are able to decrease TOC to 28 mg/L, TN to 5 mg/L, TP to 6.80 mg/L and *E. coli* to 26,200/100 mL from the respective influent values of 93.8 mg/L, 16.6 mg/L, 9.6 mg/L and 2.6×105. On integrating the system with TiO^2 based on photocatalytic oxidation for 6 h with 10 g/L of TiO^2. *E. coli* and TOC concentrations were reduced up to 1/100 mL and 6 mg/L, respectively. Reclaimed water meets European bathing quality standards. However, further investigation should be carried out for optimal utilization of TiO^2.

Gross *et al.* (2007) developed a novel recycling VFCWs mainly in the dimension of low tech, economically sound, and easy maintains. The study's system is a combination of VFCWs and trickling filters mainly to treat greywater for households for irrigation purposes. The substrate in VFCWs is composed of 15 cm of planted organic soil as a top layer, followed by 30 cm of plastic or tuff and 5 cm of pebbles as a bottom layer. The long term and short term analysis of the study elucidate the system removal efficiencies by reducing TSS, BOD, COD, TN, TP, surfactants, boron, and fecal coliform from respective influent values of 158 ± 30 mg/L, 466 ± 66 mg/L, 839 ± 47 mg/L, 22.8 ± 1.8 mg/L, 34.3 ± 2.6 mg/L, 4.7–15.6 mg/L, 1.6 ± 0.1 mg/L, $5 \times 107 \pm 2 \times 107$ /100 mL to corresponding effluent values of 3 ± 1 mg/L, 0.7 ± 0.3 mg/L, 157 ± 62 mg/L, 6.6 ± 1.1 mg/L, 10.8 ± 3.4 mg/L, 0.4–1.3 mg/L, 0.4–0.8 mg/L and $2 \times 105 \pm 1 \times 105$ /100 mL to the corresponding effluent values. The study's cost analysis showed a return over investment (ROI) to be within three years approximately.

In the study of Kadewa *et al.* (2010), a comparative analysis was performed between a novel unplanted cascading filter with planted VFCWs. VFCWs planted *Phragmites australis*, a mixture of sand, soil, and organic matter as media depicted removal efficiencies of 81.9% of turbidity, 96.12% of BOD, and 93.9% of COD from the respective influent values of 17.7 mg/L, 43.9 mg/L and 151 mg/L, which were on par with cascading filter system. However, better removal efficiency of 75.5% from the respective influent surfactant concentration of 1.39 mg/L is observed for planted VFCWs in comparison with cascading filter removal efficiency of 37.4%.

Ramprasad & Philip (2016) conducted experiments to understand the potential of VF CWs in treating greywater emerging from student hostel enriched with surfactants and personal care products compared to HF CWs. In the study (Ramprasad & Philip 2016), both the wetlands were planted with *Phragmites australis* monitored for one year with propylene glycol (PG), sodium dodecyl sulfate (SDS), and trimethylamine (TMA) as selected primary targeted elements. The study results manifested that VFCW's efficiency in treating PG, SDS, and TMA is 95%, 89% and 98%, respectively is marginally high compared to the respective removal efficiency of HFCWs. The removal efficiency of COD, BOD, TSS, TP and TN are 95%, 91.25%, 88.57%, 98.4% and 98.7% respectively from the corresponding influent 160 mg/L, 64 mg/L, 140 mg/L, 2.47 mg/L, and 18.1 mg/L. The reclaimed water from both setups meets the reuse guidelines of USEPA.

Hybrid constructed wetlands

A hybrid system combined with HFCW, followed by VFCW, is examined by Paulo *et al.* (2013) for treating mixed greywater emerging from Brazil's households. In the study (Paulo *et al.* 2013), HF SSF CW is designed with a mixture of fine and coarse gravel as media which remove 20% of TN and 80% BOD and immediately effluent from HFCW is fed intermittently to VFCW with fine mixture sand, fine gravel, and coarse gravel as media to remove the remaining BOD and TN. The substantial removal efficiency of 95%, 88%, 95%, 58%, 82% and 98% for respective parameters turbidity, COD, BOD, TP, TN and total coliform for the corresponding influent of 254 NTU, 646 mg/L, 435 mg/L, 5.6 mg/L, 8.8 mg/L and 5.4×108/100 mL were noticed.

Jokerst *et al.* (2011) examined the potential hybrid system of wetlands in treating greywater over a period of a year. In the study (Jokerst *et al.* 2011), FWS CW is planted with *Typha latifolia* which was followed by SSF HFCW vegetated by *Scirpusacutus*; FWS CW is mainly used as a preliminary treatment to trap coarse material. The observed removal efficiencies for the treatment unit during the summer season are 93.04%, 92.8%, 94.07%, 80% and 98.7% of BOD, surfactants, TN, TP and *E. coli* respectively for the corresponding influent 86.3 mg/L, 2.63 mg/L, 13.5 mg/L, 4 mg/L and 543 /100 mL. Hybrid system performance in summer is on balance with fall and spring. However, it is sub standardized during winter with a recorded removal efficiency of 47.2, 74.75, 47.47, 47.5 and 24.8% for respective parameters of BOD, surfactants, TN, TP, and *E. coli*.

To understand the precipitation effect on removal efficiency of the BOD in a hybrid system, Paulo *et al.* (2013) experimented a hybrid constructed wetland for treating greywater. The study's treatment unit (Paulo *et al.* 2013) consists of a grease trap, sedimentation tank followed by HF SSF CW with VFCW. Global removal efficiencies of 89%, 92%, 98%, 98% and 99% of COD removal were noticed for the corresponding precipitation values 2.62 ± 8.44 mm/day, 0.07 ± 0.34 mm/day,

1.60 ± 5.33 mm/day, 4.21 ± 9.49 mm/day and 4.07 ± 9.13 mm/day at a respective HLR vales of 0.2,0.3,0.2,0.3 and 0.3 ($m^3\,m^{-2}\,d^{-1}$).

To improve the performance of CWs during the variable loading regime, Comino *et al.* (2013) proposed a novel recycling hybrid CWs to treat the greywater. The study's treatment unit (Comino *et al.* 2013) is equipped with two units of VFCWs operating in parallel mechanisms, followed by HF SSF CW. The performance of the system was evaluated at different loading rates and different vegetative conditions. During the operation at optimal flow and standard pollutant load with no vegetation, the treatment unit showed an overall removal efficiency of 75.8% for COD, whereas when operated with double, triple, and quadruple times the standard pollutant load and at the standard design flow rate, treatment unit manifested a removal efficiency of 93.1%, 95%, and 95% of COD, respectively. However, by increasing the flow rates to 10 times the design flow rate, the expected removal efficiencies were not appreciable. On the other hand, almost 100% removal COD is obtained for the vegetated system at all the pollutant loads.

Green walls and green roof

Avery *et al.* (2007) develop a novel green roof water recycling system (GROW) for treating low strength greywater. The study novel GROW system planted with multiple plants is compared with HFSSF CW, and VFCWs vegetated with *Phragmites australis*. The removal efficiency of organic matter in GROW is on par with HFSSF and VFCWs, nevertheless, GROW treatment exhibited better performance in the removal of turbidity and suspended solids in comparison to the conventional CWs.

Fowdar *et al.* (2017), for the first time, examined various operating parameters and design parameters like loading rates, different types of vegetation, inflow concentration, filling material, and modified saturated zone effecting the performance of green walls in treating greywater. The study results manifested prolific removal efficiency mainly for plant species like *Canna lilies* and *Carex appressa* in removing organic matter and suspended solids, but limits in nutrients removal remarkably few species of the plant shown negative trends in nutrient removal. To analyze the above short comings, Fowdar *et al.* (2017) conducted experiments by introducing a carbon source (15 Urea) to promote denitrification in the media. The modified system depicted good removal efficiency for organic matter but falls shorts again in terms of nutrient removal.

To analyze the effect of different growth media on the performance of the green walls, Prodanovic *et al.* (2017) conducted various experiments. In the study (Prodanovic *et al.* 2017), river sand, vermiculite, coco coir, rock wool, fyto-foam, grow stone, expanded clay, and perlite are respective substrates used without vegetation to understand media mechanism in pollutant removal. Results of the study showed that delineated coco coir and perlite are incredibly useful in reducing organic and nutrient content. In the extension of research (Prodanovic *et al.* 2017) performed a study to understand the effect of the mixture of perlite and coir on the removal of the pollutants. The mixture with relatively high percentages of coir resulted in the slow drainage, thereby enhancing the contaminates' biological removal. On the other hand, composite with high perlite content retains the water for a short duration, resulting in a dominant physical removal mechanism.

To investigate the GROW system's adaptability in different climatic conditions, flow rates, and hydraulic loading rates, Ramprasad *et al.* (2017) conducted studies over 17 months in treating greywater generated from a college hostel. In the study (Ramprasad *et al.* 2017), GROW system was operated mainly in four stages (1. Startup stage, 2. Seasonal variation stage, 3. Flow rate variation, 4. Organic load variation). In all the phases of operation, the average BOD and COD concentration in the effluent is continuously less than 10 mg/L and 20 mg/L, respectively, appreciable high removal efficiencies are observed of organic matter during summer. TSS removal efficiency of 85–90% is noticed during all the operation phases. A maximum average removal percentage of 99% and 92% were noticed for nitrogen and phosphorous compounds during the summer seasonal regime (Table 2).

Figure 3 shows the resulting relationships of water quality indicators such as BOD, COD, *E. coli*, ammonia, TSS, and turbidity as an HRT function (Arden & Ma 2018). The plot between the following variables indicates little or no significant interrelationship. Therefore, the variation can also be based on the influence of various other factors, such as the GW's temperature and strength. Hence the correlation of many factors is detrimental to the treatment performance.

As indicated in Figure 3(a), BOD achieved the highest removal of 95% in VFSW at an HRT of 14 h. The average removal rates for BOD were obtained as 87.5, 89.8, 65.53, and 42.6% for HSSF, VFSW, GROW, and FTW respectively. The VFSW system continues to achieve the maximum probably due to continuous circulation leading to oxygenation. Constructed wetlands being a biological process, it was observed that COD removal rates were much lower compared to BOD. COD removal% using FTW was consistently low due to the floating roots' insufficient filtration of particulate organics. Various mechanisms, such as filtration, adsorption, and oxidation, aids in removing suspended solids from the system

Table 2 | Summary of constructed wetlands in greywater treatment (Arden & Ma 2018)

WQ Parameters	Physical parameters		Chemical parameters			Microbial parameters		
Units	TSS mg/L	Turbidity NTU	BOD mg/L	TN mg/L	TP mg/L	T.C log$_{10}$	F.C log$_{10}$	*E. coli* log$_{10}$
GRW								
N	2	2	2	0	0	2	0	2
Inflow	61	44	92	–	–	6.4	–	3.3
Outflow	12	15	41	–	–	3.8	–	0.85
% removed	84	77	71	–	–	2.6	–	2.4
HSSF								
N	5	4	17	3	3	0	2	1
Inflow	52	89	196	7.2	2.7	–	5.7	4.7
Outflow	21	38	25	4	2.3	–	3.9	0.1
% removed	64	47	87	44	24	–	1.9	4.6
RVF								
N	4	5	6	0	1	3	0	3
Inflow	66	65	198	–	1.9	7	–	4.3
Outflow	2	3.9	2.2	–	0.5	4.2	–	2.1
% removed	98	97	98	–	74	2.8	–	2
VF								
N	4	3	4	2	1	2	3	2
Inflow	58	67	99	4.9	5.2	8.2	5.7	3.5
Outflow	17	12	10	2.6	2.3	6.1	2.8	3.8
% removed	71	77	85	46	55	2	2.9	1.5

(United States Environmental Protection Agency (USEPA) 2000). The trend of decreasing TSS as a function of increasing HRT was observed in VFSW and FTW. As in Figure 3(f), none of the systems achieved a removal rate as per USEPA guidelines for unrestricted reuse in the case of turbidity removal. The removal percentage was high in the case of VFSW compared to other systems. A much higher removal of 97% was observed in RVF (Paulo *et al.* 2013). The ammonia removal percentage varied based on the amount of phosphorous linked to the plant biomass. As oxygenation was higher in the case of VFSW, biomass formation increased, which led to a maximum ammonia removal rate of 89%. Based on the trend observed in Figure 3(c), *E. coli* removal fluctuations cannot be taken as a predictable indicator for producing an effluent for reuse.

FUTURE PERSPECTIVE

It is important to conduct comprehensive studies of greywater treatment, which can assess and analyze the impact on future development.

- Pretend and evaluate by various scenarios for treating greywater in an open environment.
- CWs operation is often complicated and challenging, therefore, skilled labour is needed to avoid short circuiting, hence, advance technology should be designed with an auto intelligence system.
- Advanced GW treatment studies are required to optimize results based on influent water quality, turbidity, water stress, loading, and chemical requirements to set regulation standards to control the quality of treated for various usage purposes.
- Most of the studies regarding CWs are done across the globe with different climatic conditions, hence issues such as evapotranspiration must be considered while optimizing the data.
- Moreover, studies related to biomass growth regarding various wetlands are minimal; hence, such research can improve working conditions within wetlands for improved quality of treated water.

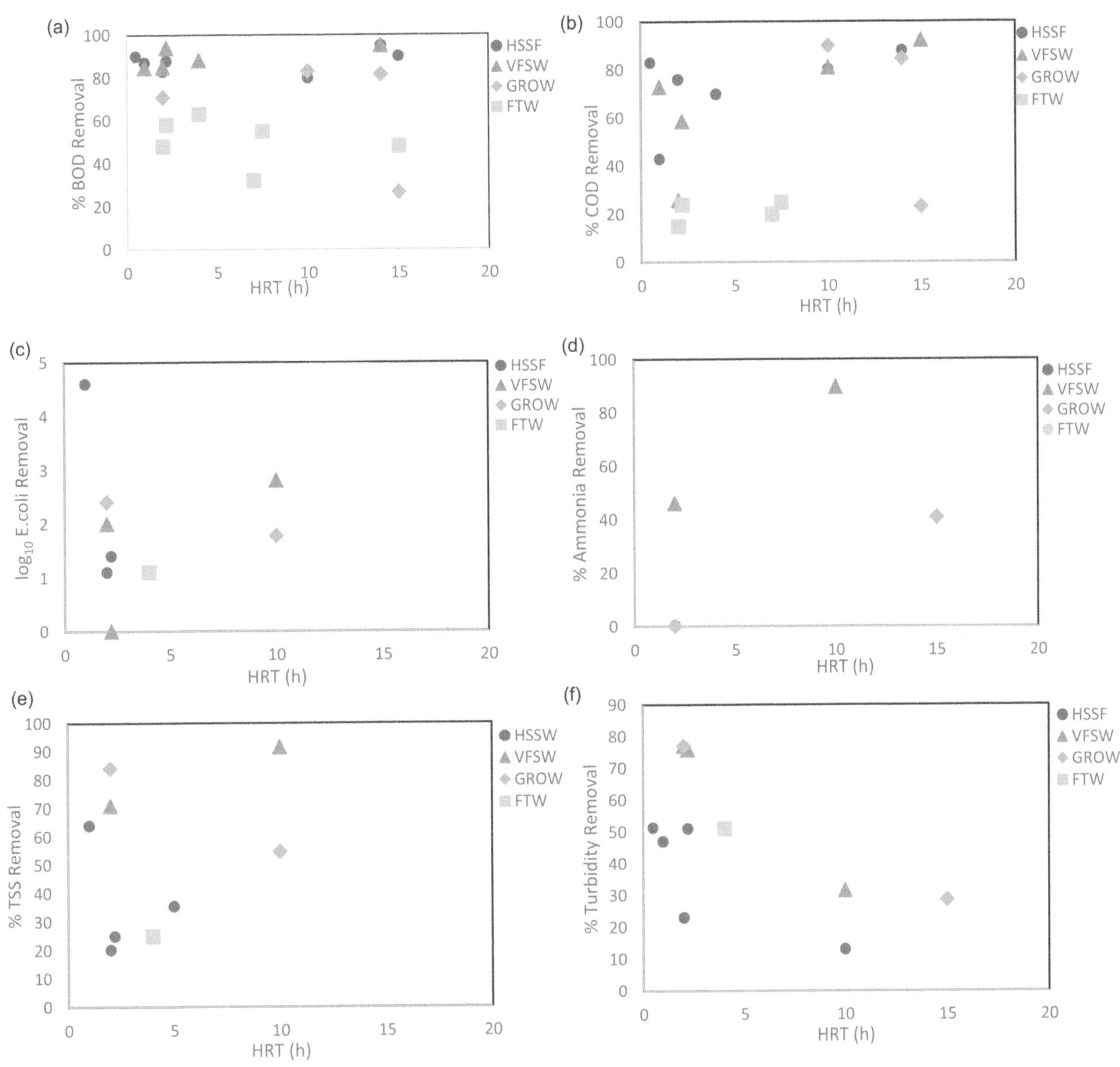

Figure 3 | Relationship between (a) % BOD removal (b) % COD removal (c) log$_{10}$ *E. coli* removal (d) % ammonia removal (e) % TSS removal and (f) % turbidity removal as a function of HRT.

CONCLUSION

Greywater makes it an attractive option in water recycling projects because of its low pollution and high volume. The average developed country produces a large amount of gravel compared to developing countries. It is noteworthy that recycling greywater can easily reduce water pressure by 30–40%, however, reusing greywater without proper purification has been shown to have negative effects on both human and animal/plant life. Built-in wetlands offer a low-cost and easy-to-operate solution to purify gravel. Typically designed wetlands operate primarily in the form of secondary or tertiary and organic matter-reducing cable and preferably suspended solids. Hybrid systems show excellent removal capacity for nutrient enrichment. To alleviate the space crisis of the swamps, new purification systems such as green roofs and green walls were developed. CWs should be optimally integrated with the appropriate disinfectant option to further reduce TC and FC in the crucible to desirable limits.

ACKNOWLEDGEMENTS

Authors are thankful to CSIR India and ICWR (International Consortium for Water Researchers) for giving support to this work through CSIR-Summer Research Training Programme 2020, by CSIR-NIEST, Jorhate.

DATA AVAILABILITY STATEMENT

All relevant data are included in the paper or its Supplementary Information.

REFERENCES

Abdel-Shafy, H. I., El-Khateeb, M. A., Regelsberger, M., El-Sheikh, R. & Shehata, M. 2009 Integrated system for the treatment of blackwater and greywater via UASB and constructed wetland in Egypt. *Desalination and Water Treatment* **8**, 272–278.

Abed, S. N., Almuktar, S. A. & Scholz, M. 2020 Impact of storage time on characteristics of synthetic greywater for two different pollutant strengths to be treated or recycled. *Water, Air, & Soil Pollution* **231**, 1–18.

Al-Hamaiedeh, H. & Bino, M. 2010 Effect of treated grey water reuse in irrigation on soil and plants. *Desalination* **256**, 115–119.

Ali, M. 2018 *Designing, Construction and Characterization of Field Scale Constructed Wetland for the Treatment of Domestic Wastewater.* Higher Education Commission, Pakistan.

Arden, S. & Ma, X. 2018 Constructed wetlands for greywater recycle and reuse: a review. *Science of the Total Environment* **630**, 587–599.

Avery, L. M., Frazer-Williams, R. A. D., Winward, G., Shirley-Smith, C., Liu, S., Memon, F. A. & Jefferson, B. 2007 Constructed wetlands for grey water treatment. *Ecohydrology & Hydrobiology* **7**, 191–200.

Berndtsson, J. C. 2010 Green roof performance towards management of runoff water quantity and quality: a review. *Ecological Engineering* **36**, 351–360.

Boyjoo, Y., Pareek, V. K. & Ang, M. 2013 A review of greywater characteristics and treatment processes. *Water Science and Technology* **67**, 1403–1424.

Comino, E., Riggio, V. A. & Rosso, M. 2013 Constructed wetland treatment of agricultural effluent from an anaerobic digester. *Ecological Engineering* **54**, 165–172.

De Gisi, S., Casella, P., Notarnicola, M. & Farina, R. 2016 Grey water in buildings: a mini-review of guidelines, technologies and case studies. *Civil Engineering and Environmental Systems* **33**, 35–54.

Dorney, J., Savage, R., Tiner, R. W. & Adamus, P., eds. 2018 *Wetland and Stream Rapid Assessments: Development, Validation, and Application.* Academic Press, Cambridge, MA, USA.

El Hamouri, B., Bey, I., Ait Douch, A., Ghazi, N. & Regelsberger, M. 2008 Greywater treatment and recycling for toilet flushing: comparison of low and high tech treatment approaches. *Water Practice and Technology* **3** (2), wpt2008041.

Elhegazy, H. & Eid, M. M. 2020 A state-of-the-art-review on grey water management: a survey from 2000 to 2020 s. *Water Science and Technology* **82** (12), 2786–2797.

Eriksson, E., Auffarth, K., Henze, M. & Ledin, A. 2002 Characteristics of grey wastewater. *Urban Water* **4**, 85–104.

Eriksson, E., Andersen, H. R., Madsen, T. S. & Ledin, A. 2009 Greywater pollution variability and loadings. *Ecological Engineering* **35**, 661–669.

Fowdar, H. S., Hatt, B. E., Breen, P., Cook, P. L. M. & Deletic, A. 2017 Designing living walls for greywater treatment. *Water Research* **110**, 218–232.

Francis, L. F. M. & Jensen, M. B. 2017 Benefits of green roofs: a systematic review of the evidence for three ecosystem services. *Urban Forestry & Urban Greening* **28**, 167–176.

Friedler, E. & Hadari, M. 2006 Economic feasibility of on-site greywater reuse in multi-storey buildings. *Desalination* **190** (1–3), 221–234.

Garcia, J., Rousseau, D. P. L., Morato, J., Lesage, E. L. S., Matamoros, V. & Bayona, J. M. 2010 Contaminant removal processes in subsurface-flow constructed wetlands: a review. *Critical Reviews in Environmental Science and Technology* **40**, 561–661.

Ghaitidak, D. M. & Yadav, K. D. 2013 Characteristics and treatment of greywater – a review. *Environmental Science and Pollution Research* **20**, 2795–2809.

Griffiths, L. N. & Mitsch, W. J. 2017 Removal of nutrients from urban stormwater runoff by storm-pulsed and seasonally pulsed created wetlands in the subtropics. *Ecological Engineering* **108**, 414–424.

Gross, A., Shmueli, O., Ronen, Z. & Raveh, E. 2007 Recycled vertical flow constructed wetland (RVFCW) – a novel method of recycling greywater for irrigation in small communities and households. *Chemosphere* **66** (5), 916–923.

Hernández-Leal, L., Temmink, H., Zeeman, G. & Buisman, C. J. N. 2011 Removal of micropollutants from aerobically treated grey water via ozone and activated carbon. *Water Research* **45**, 2887–2896.

Jasper, J. T., Nguyen, M. T., Jones, Z. L., Ismail, N. S., Sedlak, D. L., Sharp, J. O., Luthy, R. G., Horne, A. J. & Nelson, K. L. 2013 Unit process wetlands for removal of trace organic contaminants and pathogens from municipal wastewater effluents. *Environmental Engineering Science* **30**, 421–436.

Jokerst, A., Sharvelle, S. E., Hollowed, M. E. & Roesner, L. A. 2011 Seasonal performance of an outdoor constructed wetland for graywater treatment in a temperate climate. *Water Environment Research* **83**, 2187–2198.

Kadewa, W., Le Corre, K., Pidou, M., Jeffrey, P. J. & Jefferson, B. 2010 Comparison of grey water treatment performance by a cascading sand filter and a constructed wetland. *Water Science and Technology* **62**, 1471–1478.

Kadlec, R. H. & Wallace, S. 2008 *Treatment Wetlands*. CRC Press, Boca Raton, FL, USA.

Karpiscak, M. M., Foster, K. E. & Schmidt, N. 1990 Residential water conservation: Casa del agua 1. *JAWRA Journal of the American Water Resources Association* **26** (6), 939–948.

Keddy, P. A. 2010 *Wetland Ecology: Principles and Conservation*. Cambridge University Press, Cambridge, UK.

Kotti, I. P., Gikas, G. D. & Tsihrintzis, V. A. 2010 Effect of operational and design parameters on removal efficiency of pilot-scale FWS constructed wetlands and comparison with HSF systems. *Ecological Engineering* **36**, 862–875.

Kounina, A., Margni, M., Bayart, J.-B., Boulay, A.-M., Berger, M., Bulle, C., Frischknecht, R., Koehler, A., i Canals, L. M., Motoshita, M., Núñez, M., Peters, G., Pfister, S., Ridoutt, B., van Zelm, R., Verones, F. & Humbert, S. 2013 Review of methods addressing freshwater use in life cycle inventory and impact assessment. *The International Journal of Life Cycle Assessment* **18**, 707–721.

Laaffat, J., Ouazzani, N. & Mandi, L. 2015 The evaluation of potential purification of a horizontal subsurface flow constructed wetland treating greywater in semi-arid environment. *Process Safety and Environmental Protection* **95**, 86–92.

Li, Z., Gulyas, H., Jahn, M., Gajurel, D. R. & Otterpohl, V. 2004 Greywater treatment by constructed wetlands in combination with TiO$_2$-based photocatalytic oxidation for suburban and rural areas without sewer system. *Water Science and Technology* **48**, 101–106.

Lucke, T., Walker, C. & Beecham, S. 2019 Experimental designs of field-based constructed floating wetland studies: a review. *Science of the Total Environment* **660**, 199–208.

Mandal, D., Labhasetwar, P., Dhone, S., Dubey, A. S., Shinde, G. & Wate, S. 2011 Water conservation due to greywater treatment and reuse in urban setting with specific context to developing countries. *Resources, Conservation and Recycling* **55**, 356–361.

Manso, M. & Castro-Gomes, J. 2015 Green wall systems: a review of their characteristics. *Renewable and Sustainable Energy Reviews* **41**, 863–871.

Manso, M., Teotónio, I., Silva, C. M. & Cruz, C. O. 2021 Green roof and green wall benefits and costs: a review of the quantitative evidence. *Renewable and Sustainable Energy Reviews* **135**, 110111.

Mars, R., Taplin, R., Ho, G. & Mathew, K. 2003 Greywater treatment with the submergent Triglochin huegelii – a comparison between surface and subsurface systems. *Ecological Engineering* **20**, 147–156.

Matos, C., Pereira, S., Amorim, E. V., Bentes, I. & Briga-Sá, A. 2014 Wastewater and greywater reuse on irrigation in centralized and decentralized systems – an integrated approach on water quality, energy consumption and CO2 emissions. *Science of the Total Environment* **493**, 463–471.

Mehta, P. & Nagabhatla, N. 2017 *Without Water, Nothing is Secure*. UNU-INWEH Policy Brief.

Morel, A. & Diener, S. 2006 Grey water management in low and middle-income countries. Water and sanitation in developing countries (Sandec). Eawag, Switzerland: Swiss Federal institute of Aquatic Science and Technology.

Muga, H. E. & Mihelcic, J. R. 2008 Sustainability of wastewater treatment technologies. *Journal of Environmental Management* **88** (3), 437–447. https://doi.org/10.1016/j.jenvman.2007.03.008.

Oh, K. S., Cheng Leong, J. Y., Poh, P. E., Chong, M. N. & Lau, E. V. 2018 A review of greywater recycling related issues: challenges and future prospects in Malaysia. *Journal of Cleaner Production* **171**, 17–29.

Oteng-Peprah, M., Acheampong, M. A. & DeVries, N. K. 2018 Greywater characteristics, treatment systems, reuse strategies and user perception – a review. *Water, Air, & Soil Pollution* **229**, 255.

Paulo, P. L., Azevedo, C., Begosso, L., Galbiati, A. F. & Boncz, M. A. 2013 Natural systems treating greywater and blackwater on-site: integrating treatment, reuse and landscaping. *Ecological Engineering* **50**, 95–100.

Penn, R., Hadari, M. & Friedler, E. 2012 Evaluation of the effects of greywater reuse on domestic wastewater quality and quantity. *Urban Water Journal* **9**, 137–148.

Pinto, U. & Maheshwari, B. L. 2015 Sustainable graywater reuse for residential landscape irrigation–a critical review. *Chinese Journal of Population Resources and Environment* **13**, 250–264.

Pradhan, S., Al-Ghamdi, S. G. & Mackey, H. R. 2019 Greywater recycling in buildings using living walls and green roofs: a review of the applicability and challenges. *Science of The Total Environment* **652**, 330–344.

Prodanovic, V., Hatt, B., McCarthy, D., Zhang, K. & Deletic, A. 2017 Green walls for greywater reuse: understanding the role of media on pollutant removal. *Ecological Engineering* **102**, 625–635.

Pugh, T. A., MacKenzie, A. R., Whyatt, J. D. & Hewitt, C. N. 2012 Effectiveness of green infrastructure for improvement of air quality in urban street canyons. *Environmental Science & Technology* **46**, 7692–7699.

Ramprasad, C. & Philip, L. 2016 Surfactants and personal care products removal in pilot scale horizontal and vertical flow constructed wetlands while treating greywater. *Chemical Engineering Journal* **284**, 458–468.

Ramprasad, C., Smith, C. S., Memon, F. A. & Philip, L. 2017 Removal of chemical and microbial contaminants from greywater using a novel constructed wetland: GROW. *Ecological Engineering* **106**, 55–65.

Redclift, M. 2009 The environment and carbon dependence: landscapes of sustainability and materiality. *Current Sociology* **57**, 369–387.

Saumya, S., Akansha, S., Rinaldo, J., Jayasri, M. A. & Suthindhiran, K. 2015 Construction and evaluation of prototype subsurface flow wetland planted with Heliconia angusta for the treatment of synthetic greywater. *Journal of Cleaner Production* **91**, 235–240.

Schulz, C., Whitney, B. S., Rossetto, O. C., Neves, D. M., Crabb, L., de Oliveira, E. C., Terra Lima, P. L., Afzal, M., Laing, A. F., de Souza Fernandes, L. C., da Silva, C. A., Steinke, V. A., Steinke, E. T. & Saito, C. H. 2019 Physical, ecological and human dimensions of environmental change in Brazil's Pantanal wetland: synthesis and research agenda. *Science of the Total Environment* **687**, 1011–1027.

Shafiquzzaman, M., Alharbi, S. K., Haider, H., AlSaleem, S. S. & Ghumman, A. R. 2020 Development and evaluation of treatment options for recycling ablution greywater. *International Journal of Environmental Science and Technology* **17**, 1225–1238.

Shahid, M. J., Arslan, M., Ali, S., Siddique, M. & Afzal, M. 2018 Floating wetlands: a sustainable tool for wastewater treatment. *CLEAN–Soil, Air, Water* **46**, 1800120.

Sheng, X., Qiu, S., Xu, F., Shi, J., Song, X., Yu, Q., Liu, R. & Chen, L. 2020 Management of rural domestic wastewater in a city of Yangtze delta region: performance and remaining challenges. *Bioresource Technology Reports*, 100507.

Spangler, J. T., Sample, D. J., Fox, L. J., Albano, J. P. & White, S. A. 2019 Assessing nitrogen and phosphorus removal potential of five plant species in floating treatment wetlands receiving simulated nursery runoff. *Environmental Science and Pollution Research* **26** (6), 5751–5768.

Stefanakis, A., Akratos, C. S. & Tsihrintzis, V. A. 2014 *Vertical Flow Constructed Wetlands: Eco-Engineering Systems for Wastewater and Sludge Treatment.* Newnes, Oxford, UK.

United States Environmental Protection Agency (USEPA) 2000 Action Plan for Reducing, Mitigating, and Controlling Hypoxia in the Northern Gulf of Mexico (2001). USEPA, Washington, DC, USA.

Vergeles, Y., Vystavna, Y., Ishchenko , A., Rybalka, I., Marchand, L. & Stolberg, F. 2015 Assessment of treatment efficiency of constructed wetlands in East Ukraine. *Ecological Engineering* **83**, 159–168.

Vuppaladadiyam, A. K., Merayo, N., Blanco, A., Hou, J., Dionysiou, D. D. & Zhao, M. 2018 Simulation study on comparison of algal treatment to conventional biological processes for greywater treatment. *Algal Research* **35**, 106–114.

Vymazal, J. 2013 Emergent plants used in free water surface constructed wetlands: a review. *Ecological Engineering* **61**, 582–592.

Walker, S. 2017 *Design for Life: Creating Meaning in a Distracted World.* Taylor & Francis, Oxfordshire, UK.

Wang, M., Zhang, D. Q., Dong, J. W. & Tan, S. K. 2017 The AME2016 atomic mass evaluation (II). Tables, graphs and references. *Chinese Physics C* **41**, 030003.

Wang, Y., Cai, Z., Sheng, S., Pan, F., Chen, F. & Fu, J. 2020 Comprehensive evaluation of substrate materials for contaminants removal in constructed wetlands. *Science of The Total Environment* **701**, 134736.

Xu, L., Yang, S., Zhang, Y., Jin, Z., Huang, X., Bei, K., Zhao, M., Kong, H. & Zheng, X. 2020 A hydroponic green roof system for rainwater collection and greywater treatment. *Journal of Cleaner Production* **261**, 121132.

Zhang, D.-Q., Jinadasa, K. B. S. N., Gersberg, R. M., Liu, Y., Tan, S. K. & Ng, W. J. 2015 Application of constructed wetlands for wastewater treatment in tropical and subtropical regions (2000–2013). *Journal of Environmental Sciences* **30**, 30–46.

First received 4 February 2022; accepted in revised form 2 May 2022. Available online 11 May 2022

doi: 10.2166/wst.2022.149

Efficacy of *Juncus maritimus* floating treatment saltmarsh as anti-contamination barrier for saltwater aquaculture pollution control

D. Cicero-Fernandez ⬨*, J. A. Expósito-Camargo and M. Peña-Fernandez

Asociación RIA, Oficina 210, Centro Municipal de Empresas, Polígono de Trascueto, Revilla de Camargo 39600, Spain
*Corresponding author. E-mail: contacto@asociacionria.com

DC, 0000-0002-6522-8183

ABSTRACT

Floating treatment saltmarsh (FTS) is a new concept proposed to name floating treatment wetlands made of estuarine halophytes especially engineered for the control of contamination in brackish and saline waterbodies. The first full-scale FTS was implemented in 2018 to create an anti-contamination barrier for saline aquaculture wastewater treatment in an estuarine tidal lagoon. Results of a two-year investigation validated '*Phytobatea*' modular technology for floating wetlands implementation and operation. *Juncus maritimus* crossflow FTS efficiency on main mariculture wastewater constituents' removal under low hydraulic retention time was remarkable, i.e., total phosphorus (86%), total suspended solids (82%), biochemical oxygen demand (78%), total organic carbon (55%), turbidity (53%), *Escherichia coli* (30%), and dissolved oxygen increased (19%). Key features of the native halophyte *Juncus maritimus* were determined to ensure 75–100% survival under high water salinities up to 38 g/L. A scientific literature review confirmed strategic sectors' growing interest in *Juncus maritimus* as raw material, supporting its possible cultivation as an added-value by-product within integrated aquaculture systems. Plants' root systems colonization by crabs, shrimps, and young individuals of the critically endangered European eel (*Anguilla anguilla*), revealed the role of FTS for biodiversity conservation, and its potential as functional habitat, nursery, and refuge for aquatic fauna species in contaminated waterbodies.

Key words: anti-contamination barrier, constructed wetlands, floating treatment saltmarshes, halophytes, *Juncus maritimus*, saline aquaculture wastewater

HIGHLIGHTS

- Floating treatment saltmarsh (FTS) concept is firstly proposed to define saline FTWs of halophytes.
- The initial size of the plants strongly determines *J. maritimus* survival in saline contaminated waterbodies.
- Crossflow FTS are efficient solutions for mariculture wastewater treatment under high salinities and low hydraulic retention times.
- The growing interest in halophytes and their by-products supports FTS implementation within integrated aquaculture systems.
- FTS can play a strategic role as artificial habitats for biodiversity conservation and enhancement.

INTRODUCTION

Floating treatment wetlands (FTWs) are among the most innovative nature-based solutions for water contamination phytoremediation. In contrast to other wetland technologies, i.e., free water surface, vertical, and horizontal sub-surface flow treatment wetlands (TWs), plants do not necessarily require a substrate to be planted in but are directly installed in the water column through artificial buoyant structures (Pavlineri *et al.* 2017). Aerial parts of the plants grow above the water surface, and roots and rhizomes develop under water, giving shape to complex three-dimensional rhizo-filters through which water flows and is subjected to physical, chemical, and biological processes. Plants are co-responsible for pollutants remediation in synergism with biofilm microorganisms (Shahid *et al.* 2018) that grow over the surface of the wetland's submerged components, both organic and artificial (Zhang *et al.* 2016).

Buoyancy and no requirement for a substrate are the main differential features that make FTWs an especially versatile and resilient technology, enabling TWs implementation in a broad range of natural, semi-natural and anthropogenic waterbodies. FTWs can be even installed where water depth or high-water level and flow fluctuation make it impossible to implement other TWs technologies, as is the case of rivers, lakes, stormwater runoff treatment ponds, and tidal environments, i.e., salt marshes, coastal lagoons, tidal flats, and estuaries (Tanner & Headley 2011; Schwammberger *et al.* 2017, Karstens *et al.* 2021; Sharma *et al.* 2021).

FTWs efficacy has been proved in a broad range of applications, including sewage, stormwater, agriculture, aquaculture, landfill and industrial wastewater treatment, bioenergy crops cultivation, habitat creation for biodiversity conservation, and nature-based solutions for coastal defence (Pavlineri *et al.* 2017; Headley & Tondera 2019; Kourkoumpas *et al.* 2019; Ware & Callaway 2019). However, optimal design, performance and management parameters need to be better defined through more full-scale and long-term case studies development (Vymazal *et al.* 2021), particularly in certain specific fields as is the case of saltwater environments and saline wastewater treatment, on which research is almost non-existent and limited to a few, but promising, lab-scale and small sized pilot systems (Sanicola *et al.* 2019; Yajun *et al.* 2019; Calheiros *et al.* 2020; Karstens *et al.* 2021).

This paper shows what we consider the first full-scale floating treatment saltmarsh (FTS). It was designed and started in 2018 for mariculture's wastewater treatment in Ria de Tina Menor, an estuary which belongs to the *Rías Occidentales y Duna de Oyambre* Special Area of Conservation (SAC). FTS was implemented as one of the conservation actions of *Convive LIFE*, a research project funded by the European Union that addresses the problem of integrating human activities in the conservation objectives of the Natura 2000 Network in the estuaries of the region of Cantabria, in the north of Spain.

The FTS was installed as a floating barrier arranged transversely to the water flow, from side to side of a tidal artificial oblong lagoon receiving the final effluent of a mariculture company's wastewater treatment plant. It was constructed with the novel system for FTWs implementation and management named *phytobatea*, and it was planted with the native halophyte sea rush *Juncus maritimus* Lam.

The main aim of the present research was to assess *J. maritimus* FTS efficacy for saline aquaculture wastewater treatment. High water salinity up to 38 g/L compromised plants survival, and the determination of key features so that the plants could survive became a major challenge and one of the main aims of the project. Secondary goals were to test *phytobatea* technology in saline wastewater, to assess FTS role as habitat for fauna species of interest, and to undertake a preliminary evaluation of the possibility of incorporating FTSs within integrated aquaculture aystems, based on a literature review of *J. maritimus* potential as a valuable crop.

The final results of the two-year research demonstrate the feasibility of novel ecotechnologies to tackle environmental issues of major importance such as aquaculture, which according to the Food and Agriculture Organization of the United Nations, accounts for ca. 50% of the world's food fish, and could be the fastest growing food sector. It generates high volumes of wastewater, typically high in suspended solids, organic matter, and nutrients, which may produce a detrimental effect if it was discharged untreated to sensitive ecosystems, i.e., estuaries, compromising their sustainability, and that of the ecosystemic services they provide. In fact, estuaries are among the most productive, but at the same time degraded, ecosystems on the planet: over 60% of European transitional waters do not reach good ecological status (Zal *et al.* 2018) as a consequence of human overexploitation, transformation, and pollution (Lotze *et al.* 2006; Kristensen *et al.* 2021).

MATERIALS AND METHODS

Study area

The origin of the wastewater effluent is a land-based mariculture company which produces phytoplankton, young sea bream and young sea bass following organic procedures. It is settled in a 30 ha saltmarsh in the estuary of Tina Menor, where the river Nansa flows into the Cantabrian Sea. Saline wastewater is produced at an average flow of 119 m^3/h, reaching peaks of 176 m^3/h. After conventional treatment in a wastewater treatment plant (WWTP) equipped with sludge thickener and total oxidation technology, it is discharged into one end of an artificial tidal lagoon 12 m width and 95 m long, that is hydraulically connected to the open estuary through a 315 mm-diameter drainpipe located at the other end of the lagoon. The drainpipe is at 1.10 m above sea level (masl), whereas average height reached by the tides in Ria de Tina Menor is 2 masl (Fernández-Iglesias & Marquínez 2002). Consequently, water level in the lagoon varies continuously depending on tidal influence and wastewater flow, with an average water depth ranging from 40 to 90 cm in neap and spring tides respectively.

For a better understanding of the hydrology of the system, tidal regime in the lagoon was studied through a bathymetric survey and a tidal influence field assessment undertaken during a spring tide cycle, on 25 May 2017. Water levels inside and outside of the lagoon, at the final discharge point of the drainpipe in the estuary, were monitored hourly. Results indicated that most of the time (9.1 hours) water flowed from the lagoon to the estuary, and that flux was interrupted only during 3.3 hours at high tide. This result corroborated that most of the time tidal influence in the lagoon is limited by drainpipe height and capacity, and predominant water flux occurs from the lagoon to the estuary, while water entering from the estuary at high tide did not mix with that in the lagoon due to the difference of densities. Average hydraulic retention time (HRT) of 6.14 hours in the lagoon and an average water velocity of 14 m/h were estimated when water is flowing along the lagoon. In each tide cycle HRT is increased depending on the time when the water flow is interrupted by the influence of the high tide collapsing the draining pipe.

Floating treatment saltmarsh set up

In February 2018, a 66 m^2 FTS was installed within the tidal lagoon, at a distance of 50 m from the WWTP discharging point. A floating modular dock of 1.5 m width and 11 m length was put from side to side of the lagoon as a structural axis for the wetland system, and a series of 11 *Phytobatea* modules were attached to each side of the dock. The system was anchored to both lagoon banks with swinging systems, permiting FTS vertical movement with water level fluctuation.

Phytobatea technology is a modular system for FTWs implementation and management. Each module consists of a glass fibre reinforced polyester (GFRP) main support beam traversed by a plurality of GFRP rods to which plants are fixed, and an overlaying high density Polyethylene (HDPE) mesh through which plants leaves and stems grow, securing plants' verticality and facilitating biomass periodical harvesting above the water surface, whereas roots and rhizomes grow under water. Bouyancy and stability is provided by sea fishing buoys. *Phytobatea* design and structural resistance allows the operation of the modules, including their removal from the waterbody if needed. In this case, an *ad hoc* lifting machine was designed, prototyped and tested for TWs modules operation, as a tilting ramp equipped with a manual winch, which has a 200 kg load capacity, and attached with a textile sling to the dock's cleats.

The FTS infrastructure consisting of the modular dock and the vegetated Phytobateas can be considered as a cross-flow anti contamination barrier 11 m long, 7.5 m width (see Figure 1), with an under water root filter divided into two layers, a densely rooted superficial layer of 9–12 cm depth, and a less dense layer formed by longer roots of 26–36 cm length, so in neap tides, when water depth is around 40 cm, most of the water flux passes through the root systems, whereas in spring tides, when water level rises up to 90 cm, above 50% of water flows underneath the roots system. Considering the 9.1 hours during which water flows through the lagoon to the final discharge point into the estuary, and the 3.3 hours during which water flux is interrupted by the high tides collapsing the draining pipe, an average HRT of 44 minutes was estimated for water passing through the FTS at an average speed of 10.27 m/h.

Figure 1 | General view of the crossflow floating treatment saltmarsh of *Juncus maritimus* during process of installation through *Phytobatea* modules attached to a floating dock in the tidal lagoon receiving mariculture wastewater.

Plants selection, supply, installation, and maintenance

Following the same principles as in our previous research on TWs development for wastewater treatment in mountain areas (Cicero-Fernandez *et al.* 2019) plant species selection for this project was undertaken considering the importance of working with native emergent key species, thus contributing to biodiversity conservation (Doody 2010). Therefore, a preliminary study of Ria de Tina Menor saltmarsh vegetation was undertaken. It concluded that the most abundant species of native halophytes in the tidal mudflats of the estuary are *Spartina maritima*, *J. maritimus* and *Halimione portulacoides*.

H. portulacoides (*Atriplex portulacoides*) is a shrub that frequently grows in association with *J. maritimus* sharing the same biotopes, and it is a perennial high biomass producer halophyte. However, it was discarded because of its woody structure and lack of aerenchyma, which is considered a disadvantage for wastewater phytoremediation. From the two native emergent helophytes, *S. maritima* and *J. maritimus*, the second was finally selected because its root system is deeper than that of the cordgrass, which tends to grow superficially (Castillo *et al.* 2010), and it is generally accepted that FTWs' efficacy for water contamination remediation mainly depends on a vigorous root system and is related to biofilm development. Consequently, *J. maritimus* was considered the best species fitting the required features for saline wastewater phytoremediation, noting that it is a perennial halophyte, high biomass producer, which can propagate both sexually and asexually, through rhizomes and stolons. It was the unique native emergent species spontaneously present in the margins of the tidal lagoon where the FTS would be installed, indicating a probably better adaption capacity to the conditions in the site than other species. It is among the most abundant native plant species in the local saltmarshes, and a key species within the Natura 2000 Network priority habitat 1330, Atlantic salt meadows (*Glauco-Puccinellietalia maritimae*) (Doody 2010), so planting it in the FTS also contributes to local biodiversity conservation through *J. maritimus* local genes preservation, thanks to the spreading of the seeds of the plants installed in the FTS.

A scientific literature review about the interest in *J. maritimus* as raw material for different strategic sectors was undertaken, confirming the possibility of *J. maritimus* FTS implementation not only for wastewater remediation, but also as an interesting crop that could be part of an integrated aquaculture system.

With the preceptive authorisations, 440 *Juncus maritimus* plants were supplied from a nearby salt meadow. Big mats were taken from the mud using hand tools. During the washing of plants with water in order to eliminate any soil, most of them broke up into smaller mats and rhizomes. Before plantation, rushes were classified according to mat or main rhizome length above or below 10 cm, and leaves were pruned to around 20 cm long. Plants were installed at a density of 6.6 units/m^2 (20 units/*Phytobatea*), in a proportion of 50% of each rhizome size, following a traditional staggered pattern. Months later, unsuccessful plants were replaced with new individuals of varied sizes and ages, supplied from the same area, following the same protocol for preparation and installation except for leaves pruning. No mowing or any other plant maintenance operations were done during the experiment.

Monitoring

FTS was monitored on a regular basis for two years, from March 2018 to March 2020.

Plant survival was assessed every six months over a representative sample of 60 plants (13.6% of the total population) distributed in groups of 10 plants in 6 phytobateas strategically located in the 2nd, 6th, and 10th positions at each side of the floating dock. For vegetative growth evaluation, persistent organs, i.e., mat's length and width, and root systems length, were measured yearly. Green leaves and sprouts were considered the main survival indicators. Time needed for FTS initial installation (i.e., plants attachment to *Phytobatea* modules and installation in the water), as well as for dead plants replacement (i.e., modules removal, plants replacement and modules return to the water), was recorded to evaluate *Phytobatea* technology efficiency for TWs' vegetation installation, management, and maintenance.

Water was sampled monthly for 25 consecutive months, always in days with the lowest coefficient tides, therefore preventing tidal influence on and interferences with predominant water quality gradient along the system. Samples were taken in four locations within the lagoon (see Figure 2) according to water flow direction: the first one (A) was in the discharging point of the conventional wastewater treatment plant at one end of the lagoon, the second one (B) just before the FTS, in the centre of the lagoon's section; the third sampling station (C) was established just after the FTS; and the last one (D) at the other end of the lagoon, by the mouth of the drainpipe through which final effluent discharges into Tina Menor estuary. When the *Phytobatea* modules were removed from the waterbody for maintenance operations, organoleptic water properties beneath the saltmarsh appear to be better than in the rest of the lagoon in terms of transparency and odour, so from August 2019 to March

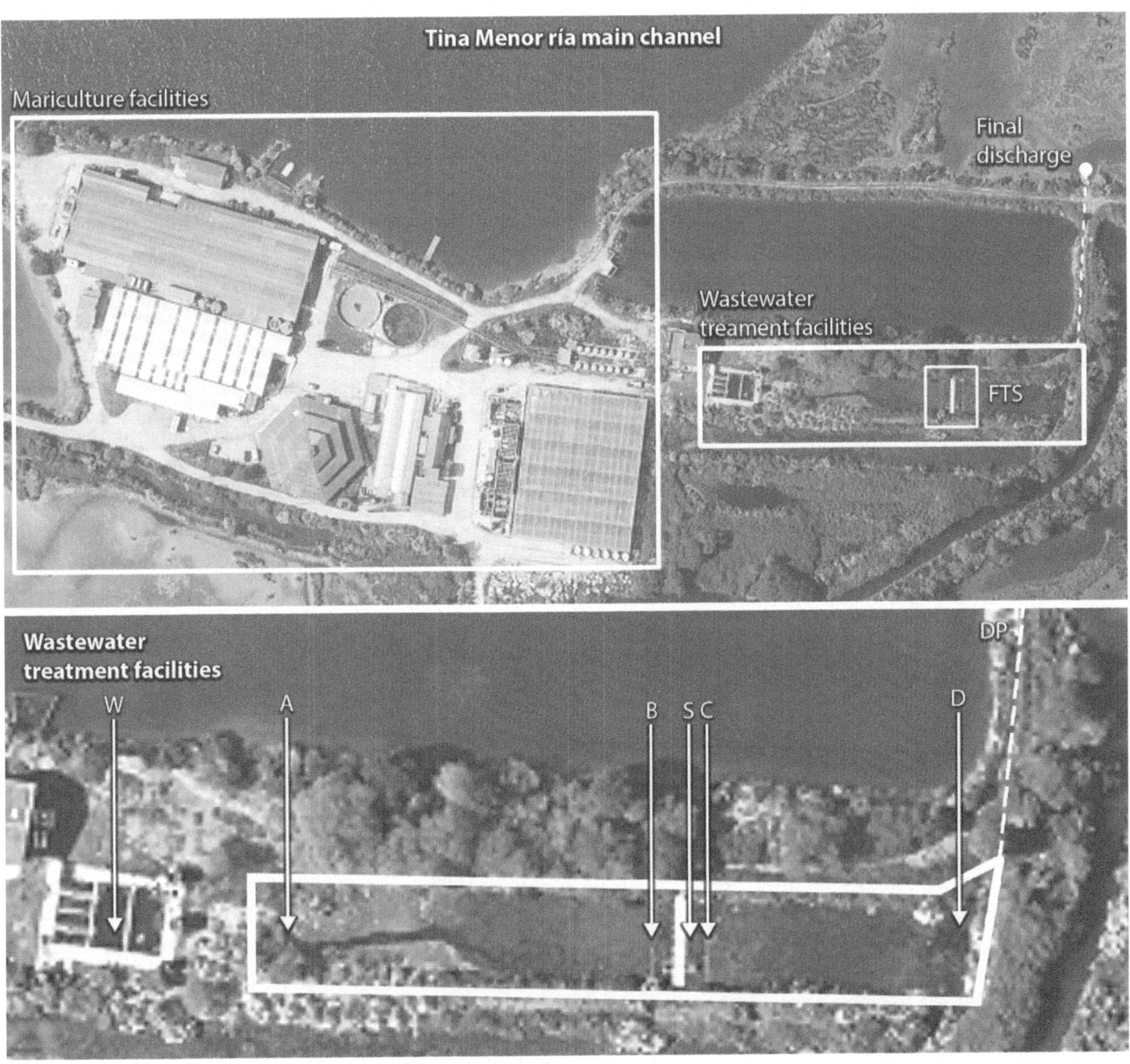

Figure 2 | Site description maps. In the upper orthophoto a general overview of the mariculture facilities is provided, with FTS location. Enlarged image below shows the location of the conventional Wastewater Treatment Plant (W) from which effluent is discharged at the beginning of the tidal lagoon, where there is the first water sampling station for water quality monitoring (A), followed by station B (just before the FTS), S (in the centre of the FTS), C (just after the FTS), and D, located by the mouth of the drainpipe (DP) through which final effluent discharges into Tina Menor estuary.

2020, an additional sampling point (S) was established in the centre of the FTS to evaluate if a substantially better water quality existed within the wetland.

On site analyses were conducted for temperature (T), pH, dissolved oxygen (DO), salinity and turbidity, whereas analyses for conductivity (Cond), biochemical oxygen demand (BOD_5), total organic carbon (TOC), total suspended solids (TSS), total nitrogen (TN), total phosphorus (TP), total coliforms (TC) and *Escherichia coli* (*E. coli*) analyses were conducted by the Centre for Environmental Research of the Government of Cantabria (CIMA) Certified Laboratories following standard methods.

When fauna was observed in the FTS during maintenance and monitoring works, those species were identified, noted, and photographed, if possible, for a preliminary evaluation of FTS role as habitat.

Data analysis

Hypothesis of the minimum size for *Juncus maritimus* survival and growth

Plant survival monitoring results suggested that the size of the plants could be the key factor for sea rush individual success. To confirm this hypothesis, a statistical study was done over a representative sample (14%) of plants from the initial plantation to determine if size of persistent parts of the plants (i.e., mat's length and width, and roots length) were related to survival and death rates. Average values of each parameter were calculated for both groups of dead and alive plants respectively. Data were analysed through one-way ANOVA, and one-sided T-Test was applied to determine if there were differences (p-value < 0.05) between the estimated average sizes of dead and alive plants. To discard significative water salinity variation influencing survival rates during the experiment, water conductivity values corresponding to high survival rate period were compared to values corresponding to low survival rate period through the T-Test. Cumulative growth assessment of the plants was conducted on those plants forming part of the monitoring sample that survived during the two-year investigation.

Floating treatment saltmarsh performance

Average concentrations of water parameters were calculated with recorded data from a total of 25 sampling campaigns for A, B, C and D sampling points, and of 7 campaigns for B, C and S sampling points, and removal/increase rates expressed in percentage were also calculated. To discriminate FTS efficacy on wastewater treatment from other processes that could occur in the lagoon (i.e., sedimentation, algae phytoremediation), removal rates from A to B and from C to D were calculated to determine removal rates in the previous and subsequent lagoon sections to the FTS and compared to those obtained from B to C sampling points, representative of the saltmarsh performance through one-sided T-Test at a 95% confidence level, and difference between sampling points was considered as statistically significant if the p-value < 0.05. In addition, water quality average values in B and C (sampling points located just before and after the TWs) were compared with S (additional sampling point located in the centre of the wetland) following the same method, to confirm if better water quality existed in the FTS centre.

RESULTS AND DISCUSSION

Floating treatment saltmarsh terms and classification

As natural wetlands, depending on water salinity TWs can be divided into freshwater and saline, which in turn comprise brackish (salinity 5–30 g/L) and saltwater (salinity > 30 g/L) wetlands (Yang 2019). At the same time, as well as freshwater ones, saline TWs can be designed following the best available technologies, i.e., free water surface, horizontal subsurface flow, vertical subsurface flow, and floating wetland systems, of which the latter is the youngest and most innovative type. This is the main reason of the range of terms used to name it. However, the term *Floating Treatment Wetland (FTW)* coined by Headley & Tanner (2008) appears to be the most widely accepted to date (Colares *et al.* 2020; Shen *et al.* 2021; Vymazal *et al.* 2021). FTWs are floating systems for water contamination control planted with emergent macrophytes. Therefore, when halophytes and/or salt tolerant shrubs, herbs or grasses are used to create any FTW, to be implemented in saline waterbodies and/or to treat saline wastewater, *Saline Floating Treatment Wetlands* would be the logical name to define it, as a result of the current definitions for saline constructed wetlands (CWs) and FTWs combination. However, instead of *Saline Floating Treatment Wetlands*, the authors of the present article consider the term *Floating Treatment Saltmarsh* (FTS) as the most synthesizing, understandable, and ecologically accurate term to define the novel type of TW which is described and evaluated in the present investigation.

Juncus maritimus survival and growth

Juncus maritimus is documented to be one of the most salt tolerant halophytes of the *Juncus* genus, capable to survive salinities up 30 g/L (Boscaiu *et al.* 2011). However, when this investigation was developed, no reference studies were found, neither on survival rates when exposed to higher salinities nor under hydroponic regime, so a trial/error process was undertaken to determine sea rush ability to overcome extreme conditions of the present investigation.

As can be seen in Figure 3, six months after initial plantation a high mortality up to 45% occurred. This was related not only to salt and chemical stress to which plants were exposed, but also to leaves pruning prior to plantation, which could compromise fundamental adaptation mechanisms of *J. maritimus* to salt stress that depend on photosynthesis, as is the case of osmolytes, i.e., sugars and proline synthesis and accumulation in shoots (Hassan *et al.* 2016). Consequently, in August 2019 dead plants were replaced with one year old individuals (mat diameter ≤10 cm) with intact leaves, transplanted

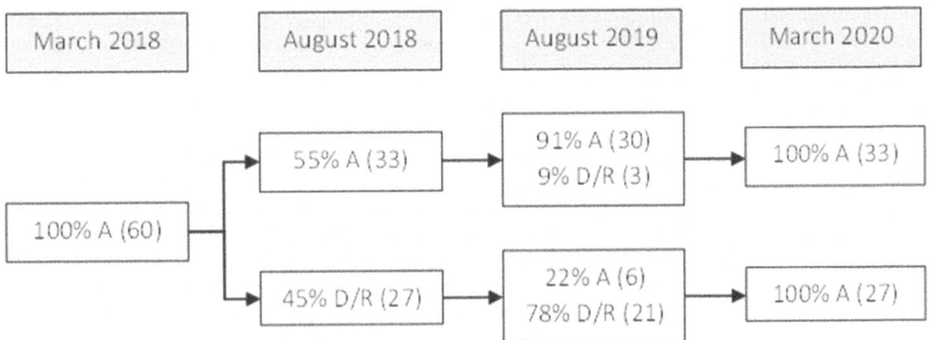

Figure 3 | Flowchart of *Juncus maritimus* plants representative sample survival assessment along 2-year monitoring, where 'A' means alive plants, and 'D/R' means dead plants that were replaced with new individuals in each monitoring campaign.

from a nearby saltmarsh. The first reaction after plantation was the progressive senescence of the leaves, and one year later 78% of those new young plants died, while 91% of larger plants which had survived since the initial plantation were still alive. Survival dependence on plant size under similar environmental conditions has been documented by other authors, as is the case of Lissner & Schierup (1997), who observed a higher mortality of *Phragmites australis* young plants compared to larger plants when cultivated in saline solutions, and by Calheiros *et al.* (2020), who in 2018, almost simultaneously to the present project development, were undertaking a research project on saltmarsh halophytes FTS systems feasibility for implementation in port marinas in Portugal, which concluded that small sized *J. maritimus* individuals could not survive the conditions of the experiment (salinity of 30–34 g/L).

Therefore, for a better understanding of our results, a comparative statistical analysis of plants' key features sizes was conducted between dead and alive plants from the first plantation. Results revealed that survival notably increased, above 75%, for plants whose mat length and width, and roots average and maximum lengths, were equal to or above 22.39, 6.52, 7.82 and 29.3 cm respectively (see Table 1). To verify this hypothesis, a second campaign for dead plants replacement was carried out in August 2019, when 24 plants with sizes above the hypothetic minimum values from which survival rate notably increased were planted. Seven months later, in March 2020, 100% survival was confirmed, suggesting that the hypothesis is correct.

To determine if water salinity varied during that last period (Sep 19–Mar 20), electrical conductivity results in sampling stations located by the FTS (B and C) were compared with those of the same period obtained in the previous year (Sep 18–Mar 19), when 40% of plants died. Slightly lower salinity was observed in the last period of the study, but T-Test confirmed that no significant differences existed between both periods (Table 2), so we concluded that plants' initial size was the main factor determining *J. maritimus* survival rate when they were planted bare rooted in contaminated sea water, and that equal or above 75% survival was achieved when the length and width of the mats, and the average length of the roots were above 22, 7 and 8 cm respectively.

As Figure 4 shows, during the first year, growth of surviving plants was in general low or non-significant, which according to previous studies (Boscaiu *et al.* 2011; Hassan *et al.* (2016) is a known response of *J. maritimus* to salt stress. In the second

Table 1 | Summary of the main data for minimum size hypothesis for *Juncus maritimus* survival to high salinity stress formulation and verification

Key features (cm)	March 2018 plantation		Aug. 2019 plantation (Survival = 100%)			Minimum size hypothesis (Survival ≥ 75%)
	Alive Av.	Dead Av.	Min.	Av.	Max.	Min
Mat/main rhizome length	22.39	19.57	28	34.67	50	22
Mat width	6.52	3.72	6	13.56	25	7
Root system av. length	7.82	6.41	12	18.56	36	8
Root system max. length	29.3	25	12	18.56	36	29

The first column of data represents average sizes of dead and alive key features of the plants recorded in August 2018, for the group of plants installed in March 2018. The second column represents average minimum (Min.), mean (Av.) and maximum (Max.) values for the main key features of sea rush individuals planted in August 2019 exceeding minimum sizes determined for a survival rate equal or above 75%, represented in the third column.

Table 2 | T-Test statistical comparison of water conductivity average values in sampling stations located by the edges of the FTS (B and C) between September 18 and March 2019 period, when a high mortality of *J. maritimus* was observed, and September 2019–March 2020 final period, when plant survival increased to 100%

Conductivity		B (mS/cm)	C (mS/cm)
September 18–March 19	Average ± s.d.	48.40 ± 2.10	48.76 ± 1.54
	Maximum	50.70	51.10
	Minimum	45	46.70
September 19–March 20	Average ± s.d.	39.74 ± 10.80	40.63 ± 10.80
	Maximum	49.00	48.70
	Minimum	20.40	20.80
Averages comparison	*P*-value (T-Test $\alpha < 0.05$)	0.051657113 Non-significant	0.05594512 Non-significant

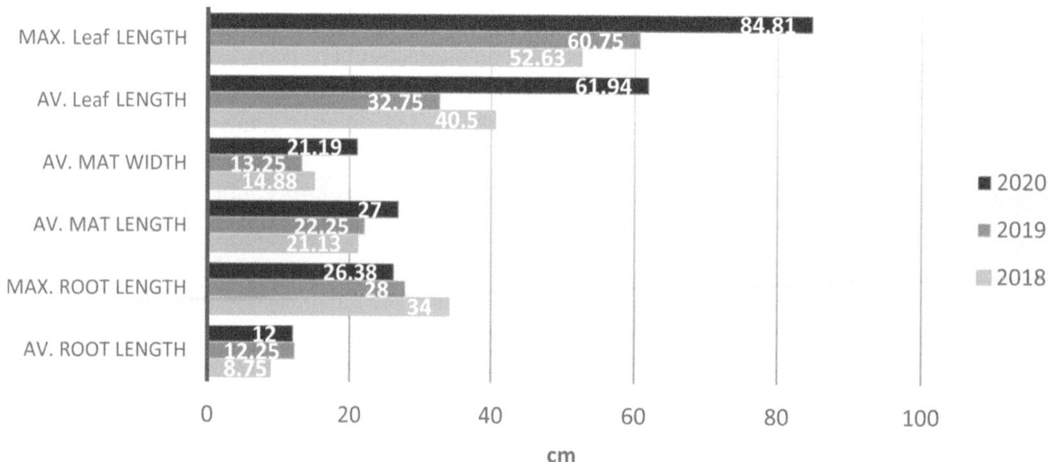

Figure 4 | *Juncus maritimus* growth results during 2-year monitoring.

year, a remarkable growth of mats and leaves of plants that were still alive was observed, while incipient rhizomes and stolons were firstly detected in the last monitoring campaign, in March 2020, indicating the plants passing from an initial phase focused on mere survival and adaptation, to an establishment and propagation phase starting in the third year after plantation, which is consistent with previous findings relating tolerance to salt stress with plant age and growth stage (Chartzoulakis & Klapaki 2000; Delattre *et al.* 2021).

FTS efficacy as anti-contamination barrier

Water quality in the tidal lagoon was mainly determined by saltwater aquaculture wastewater effluent. Despite being treated in a wastewater treatment plant prior to dumping in the tidal lagoon, the main parameters were generally above typical values for mariculture wastewater, particularly for suspended solids and nutrients, in comparison with land-based mariculture effluents quality data compiled by De Lange *et al.* (2013).

When 25 consecutive months' water quality average data from each sampling station were compared, an increasing trend of overall concentrations of constituents was observed from the beginning of the lagoon to the front side of the FTS, where most parameters under study reached their peaks (see Table 3); however, only TC and *E. coli* concentrations increments were statistically significant along the first 50 m section of the lagoon, and the same was observed in the second 37.5 m unvegetated section, downstream of the FTS, where only salinity was significantly lower. These results demonstrate that no wastewater treatment that might occur because of lagooning, algae growth, dilution, or other processes was taking place in non-vegetated sections of the water body. In contrast, high average removal rates for all parameters were observed from B to C, which is the section of the lagoon corresponding to the *J. maritimus* floating treatment saltmarsh (see Table 3). Even though the FTS

Table 3 | Columns A, B, C and D represent water quality parameters (25 months average ± standard deviation) in the discharging point of the conventional aquaculture wastewater treatment plant at the beginning of the lagoon (A), 50 m downstream, just before the FTS (B), just after the FTS (C), and at the other end of the lagoon, by the mouth of the drainpipe (D)

	A	B	C	D	% A-B	% B-C	% C-D
T (°C)	17.51 ± 3.75	17.05 ± 4.44	16.73 ± 4.39	16.56 ± 4.36	*2.63	*1.88	1.01
pH	9.36 ± 0.94	9.35 ± 0.99	9.35 ± 0.98	9.35 ± 0.99	0.11	0.00	0.00
Cond. (ms/cm)	47.14 ± 4.46	45.01 ± 6.80	45.40 ± 6.70	45.15 ± 7.34	*4.52	−0.87	0.55
Salinity (g/L)	33.48 ± 3.89	32.16 ± 5.65	31.92 ± 5.38	31.48 ± 5.72	3.94	0.75	*1.38
DO (mg/L)	2.76 ± 1.30	2.67 ± 1.56	3.18 ± 1.48	3.03 ± 1.59	3.26	* − 19.10	4.72
Turbidity	22.30 ± 25.65	36.10 ± 40.08	16.98 ± 15.98	14.75 ± 23.31	−61.88	*52.96	13.13
TSS (mg/L)	175.38 ± 471.33	<210.12 ± 370.75	37.22 ± 43.95	<54.50 ± 107.31	−19.81	*82.29	−46.43
BOD$_5$ (mgO$_2$/L)	<33.08 ± 68.77	<67.64 ± 138.65	<15.16 ± 9.35	<41.36 ± 100.16	−104.47	*77.59	−172.82
TOC (mgO$_2$/L)	<13.71 ± 11.19	<26.73 ± 39.51	<11.90 ± 6.88	<13.87 ± 25.35	−94.97	*55.48	−16.55
TN (mg/L)	6.84 ± 5.70	7.96 ± 9.20	6.17 ± 4.86	5.62 ± 6.32	−16.37	22.49	8.91
TP (mg/L)	12.15 ± 37.97	17.14 ± 36.86	2.42 ± 2.26	<6.86 ± 20.49	−41.07	*85.88	−183.47
TC (MPN/ 100 ml)	>4,193.40 ± 9,078.45	>8,502.96 ± 11,811.38	>6,171.36 ± 7,831.60	>38,488.32 ± 154,054.54	* − 102.77	27.42	−523.66
EC (MPN/ 100 ml)	<81.76 ± 202.38	<148.72 ± 250.04	<104.44 ± 208.19	<300.44 ± 700.49	* − 81.9	*29.77	−187.67

Columns on the right side of the table represent the removal or increment (−) rates of the parameters (in percentage) along unvegetated sections (A-B and C-D), and in the floating treatment saltmarsh (B-C). Statistically significant removal rates have been calculated at a 95% confidence level (p-value < 0.05) and are indicated with an asterisk (*). MPN, most probable number.

represented 7% of the total surface and 8% of the total length of the lagoon, high average removal rates were performed, with the better results for TP (86%), TSS (82%) and BOD (78%), and lower but also significant efficiencies on TOC (55%), turbidity (53%), and *E. coli* (30%) removal. In addition, a 19% average increase of dissolved oxygen was also achieved by the FTS.

The overall increment of pollutant concentrations in the anterior zone of the FTS indicated an efficient contamination containment performance of the saltmarsh, clearly visible to the naked eye, suggesting FTS's potential as anti-contamination barriers in shallow waterbodies. This application could be developed as an alternative nature-based solution to *silt curtains* commonly used for turbidity mitigation, i.e., during dredging operations, the efficacy of which is under discussion since Radermacher *et al.* (2015) demonstrated that hanging silt curtains are not effective in cross currents because the synthetic cloth of which this technology consists is not permeable to water flux, leading water to pass underneath and around the edges of the curtain creating forced currents and consequently increasing water velocity, turbulence, and thus promoting pollution dispersal. On the contrary, FTWs are permeable rhizofilters that allow water to pass through the system, thus preventing water flux diversion and boosting pollutants removal, as is the case of the FTS implemented in the present investigation, the performance of which was 52.96% average turbidity mitigation. In addition, after water treatment through the FTS, turbidity average value remained virtually steady until the end of the lagoon, indicating that no near-bed or near-shores currents were playing a significant role transporting pollution downstream, and that pollution control was efficiently and unequivocally performed by the *J. maritimus* FTS. Accordingly, an average efficacy up to 82% for TSS removal was observed, which is in line with previous studies reporting suspended solids removal performance by FTWs, which is known to be achieved mainly by physical settling and rhizofiltration processes such as trapping, coagulation, flocculation, and sedimentation (Tanner & Headley 2011; Pavlineri *et al.* 2017; Shahid *et al.* 2018). A very similar removal efficacy (i.e., 85.88%) was monitored for total phosphorus, which according to Tanner & Headley (2011) besides plant uptake, greatly depends on sorption in fine suspended solids.

Similar trends followed by both parameters from the highly variable data before the FTS (B) to the stable concentrations just after (C) can be seen in Table 3. Results up to 5- and 7-fold lower in C were calculated for TSS and TP respectively, and low standard deviations in C highlighted FTW efficacy and resilience against TP and TSS peaks, as well as TP concentration decrease related to TSS removal. Aside from this, exceptionally high TP removal results, far above the 48.75% mean rate determined by Pavlineri *et al.* (2017), could have been caused by the particularly high TP influent concentration; according to the same authors, these estimations are positively correlated with TP removal rate.

On the other hand, 22.49% TN removal efficiency was below mean rates reported in previous studies for TWs in general, and for FTWs and saline TWs in particular (Vymazal 2007; Liang *et al.* 2017; Pavlineri *et al.* 2017), which apart from aerobic and anaerobic zoning for nitrification-denitrification processes and well-developed root systems and associated microorganisms, highlighted the importance of HRT (i.e., days) on nitrogen compounds transformation and removal, which might be the main cause conditioning our results. A similar trend (i.e., 17% average TN removal) was reported by Schwammberger *et al.* (2017) for a stormwater FTW implemented in a lake intermittently receiving water runoff from an urban area. In addition, water salinity above *J. maritimus* optimal ecological range might be an additional limiting factor for TN removal, as Yajun *et al.* (2019) reported for the halophyte *Suaeda salsa* FTS when implemented in eutrophic brackish water.

Indicators of organic (BOD and TOC) and microbiological pollution (TC and *E. coli*) reached high average values in D, particularly the pathogen concentrations, which reached their maximum values at the end of the lagoon, indicating faecal contamination entering from the estuary with the tides. This occurred only at certain moments of the year, mainly in July and August, and it was probably originated by local farmers fertilising nearby agricultural fields using liquid bovine slurry, which is afterwards washed by runoff and spilled into the estuarine waterbody as non-point source pollution. However, influent contaminated water did not influence sampling stations B and C, and FTS performances on these parameters' removal could be assessed as can be seen in Table 3. BOD and TOC removal rates agreed with other types of TWs for saline wastewater treatment, suggesting that FTSs might be more efficient than subsurface flow wetlands, as HRT was substantially lower in the present investigation than in the ones compiled by Liang *et al.* (2017). No previous research was found on coliforms of saline aquaculture wastewater treatment with TWs, so the 27.42% and 29.77% elimination rates for total coliforms and *E. coli* respectively performed by the *J. maritimus* FTS are considered as the first demonstration of FTS's potential for mariculture wastewater disinfection, which could be probably easily improved through FTS dimension and/or wastewater HRT increment.

FTS interior water quality and role as aquatic fauna refuge

During the maintenance and monitoring operations undertaken in August 2019, better organoleptic properties were observed in the water beneath the floating saltmarsh, i.e., water transparency and odour. In addition, individuals of the green crab *Carcinus maenas*, and young individuals of the critically endangered species European eel (*Anguilla anguilla*) were detected attached to the *J. maritimus* roots, as well as shrimps (*Palaemon sp.*), which in addition to colonising root systems were forming schools beneath the wetland. Calheiros *et al.* (2020) also reported *Palaemon* sp. and other macrofauna species colonising a small pilot FTS of halophytes made of cork in a port marina in Portugal, while Karstens *et al.* (2021) coincided on FTS attractiveness for shrimps and particularly for young European eels, which massively colonised a FTS installed in a brackish eutrophicated coastal lagoon in the Baltic Sea.

As Table 4 shows, one of the reasons for aquatic fauna colonisation might be a better water quality observed in the centre of the *J. maritimus* FTS in comparison with the anterior and posterior edges of the wetland and the rest of the lagoon. This trend was observed for all the parameters under study but for *E. coli* and DO, which reached better results in C. The rest of the parameters were remarkably lower in S, particularly for TSS, turbidity and TP, which were up to 4-fold, 3-fold, and 2.78-fold lower in the centre of the wetland than in the edges respectively, and differences between B and S for these parameters were statistically significant according to the results of one-sided T-test. Salinity was close to 2 points lower and pH 1 point lower, indicating the formation of a good water quality mass below the FTS, that could play a major role as a refuge for aquatic fauna (see Figure 5). Similar trends were reported in the contaminated estuary of the Yangtze River by Huang *et al.* (2020), which proved the efficacy of *Phragmites australis* FTS as larvae fish assemblage structure and demonstrated the relation between the higher fish-larvae density and the better water quality and supporting structure provided by the artificial FTS. This ecological application could be especially interesting in waterbodies subjected to heavy pollution events, as is the case of collapsing coastal lagoons like the Mar Menor, on the Mediterranean coast of Spain (Ruiz *et al.* 2020).

Juncus maritimus FTS integration in aquaculture systems

Integrated aquaculture (IA) systems are those in which different species of organisms are synergically cultivated in the same water (Boyd *et al.* 2020), thus increasing the efficiency and sustainability of the process. In this case, in which fish is the main product, vegetation cultivated in the FTS for wastewater treatment could also be produced as a valuable crop, if there are markets interested in *J. maritimus*.

Table 4 | Average results and standard deviations for water parameters calculated for the period August 2019–March 2020, during which sampling station S, located in the centre of the FTS, was added, to assess if a better quality was maintained in the centre of the saltmarsh compared to the edges, represented by results on sampling stations B (upstream side of the FTS) and C (downstream side of the FTS)

	B	S	C	% B-S	% S-C
T (°C)	14.45 ± 3.22	13.74 ± 3.18	14.45 ± 3.60	4.91	−5.17
pH	9.42 ± 1.21	8.84 ± 1.09	9.36 ± 1.29	6.16	−5.88
Cond. (ms/cm)	40.46 ± 6.80	38.93 ± 10.87	41.70 ± 10.22	3.78	7.12
Salinity (g/L)	28.38 ± 5.76	26.79 ± 8.72	28.63 ± 8.33	5.6	−6.87
DO (mg/L)	3.50 ± 1.57	3.69 ± 1.81	3.96 ± 1.56	−5.43	−7.32
Turbidity	16.43 ± 7.95	5.36 ± 3.32	10.12 ± 12.68	*67.38	−88.81
TSS (mg/L)	73.38 ± 19.65	17.49 ± 13.64	37.43 ± 67.75	*76.17	−114.01
BOD$_5$ (mgO$_2$/L)	$<24.50 \pm 2.56$	$<11.29 \pm 2.63$	$<12.38 \pm 24.51$	53.92	−9.65
TOC (mgO$_2$/L)	$<11.25 \pm 0.00$	$<10.00 \pm 0.00$	$<12.75 \pm 2.82$	11.11	−27.5
TN (mg/L)	3.30 ± 1.66	2.86 ± 1.78	3.54 ± 1.45	13.33	−23.78
TP (mg/L)	3.20 ± 1.81	1.15 ± 0.59	1.82 ± 1.81	*64.06	−58.26
TC (MPN/100 ml)	$<7,722.50 \pm 1,576.46$	$4,468 \pm 8,765.35$	$<4,876.50 \pm 10,397.70$	42.14	−9.14
EC (MPN/100 ml)	$<123.13 \pm 17.92$	$<222.57 \pm 482.61$	$<56.25 \pm 265.35$	−80.76	74.73

Percentage removal or increment (−) rates between the edges and the centre of the FTS are represented in the columns on the right, and statistically significant removal rates have been calculated at a 95% confidence level (p-value < 0.05) and are indicated with an asterisk (*). MPN, most probable number.

The review of *Juncus* genera species potential for phytoremediation made by Syranidou *et al.* (2016) identified *J. maritimus* as one of the species of interest for phytoremediation, finding that most of the existing investigations are focused on heavy metal and organic persistent pollutants phytoremediation of contaminated sediments, mainly produced in Portugal (Almeida *et al.* 2006; Marques *et al.* 2011; Ribeiro *et al.* 2013; Montenegro *et al.* 2016), while publications on *J. maritimus* implementation for wastewater treatment are almost non-existent, and limited to short-term experiments, such as the one developed by Yahiaoui *et al.* (2020) for domestic wastewater treatment with vertical flow TWs.

On the other hand, the scientific literature review revealed that there are different sectors interested in *J. maritimus* as raw material to produce added-value products and services. Several recent examples have been found, such as the potential use of sea rush fibres for building insulation composites manufacture (Saghrouni *et al.* 2020), or its possible cultivation for producing biomass high in carbohydrates for bioenergy (bioethanol), as an alternative crop to others that are currently compromising human food and fresh water sources (Smichi *et al.* 2015). Sea rush antifungal properties have also attracted the attention of the agriculture sector, i.e., crude methanolic extracts obtained from *J. maritimus* rhizomes have been reported by Sahli *et al.* (2018) as an effective biobased fungicide against *Zymoseptoria tritici*, the most important pathogen of wheat, which is the most important food crop in the world. There is also remarkable interest in *J. maritimus* compounds in the pharmaceutical sector. In this context, López-Álvarez *et al.* (2012) demonstrated the possibility of producing porous silicon carbide scaffolds from *J. maritimus* for tissue engineering application, i.c., for the fabrication of bio-structures for bone tissue replacement. Sahuc *et al.* (2019) demonstrated the feasibility to produce new and low-cost direct acting antivirals against hepatitis C virus based on dehydrojuncusol, a natural compound isolated from crude extract of *J. maritimus*, while Alamer & Algaraawi (2020) demonstrated antifungal activity of methanol extract from phytoconstituents of *J. maritimus* leaves and stems, to effectively treat dermatophytosis.

Consequently, *J. maritimus* could be produced in the FTS as a product to supply one or various of the sectors of interest, thus mariculture wastewater would cease to be considered as a waste to be managed but instead as a useful culture medium high in nutrients and organic matter, and hence a valuable by-product, which would mean *J. maritimus* FTS integration at a level 4 (Boyd *et al.* 2020) within the aquaculture system.

Phytobatea technology validation in saline tidal environment

After two-years' monitoring, *Phytobatea* modular devices did not show any signal of deterioration or corrosion and buoyancy was stable. All components remained in perfect condition, and the whole structure did not suffer any apparent damage. This

Figure 5 | Green crab (*Carcinus maenas*) colonising *Juncus maritimus* roots in a *Phytobatea*.

agrees with long-term investigations on GFRP composites' durability in seawater (Idrisi *et al.* 2021), which determined that immersion for 11 years slightly affected GFRP composites when temperature was 23 °C, whereas deterioration increased with higher water temperatures. Accordingly, since average water temperature in the FTS was 17 °C, no apparent deterioration symptoms in the *Phytobatea* modules were observed, and water absorption should have not exceeded 3.7% increase in weight, while tensile strength reduction should not exceed 2% after 12 months of exposure. However, mariculture wastewater might influence GFRP component durability in the long-term, and a specific investigation would be of great interest.

Plants grew correctly attached to the fibreglass rods and through the HDPE mesh, maintaining shoots' verticality even after storms, heavy winds, and snow accumulation (see Figure 6). Monitoring of installation process revealed that average time for a FTS implementation, including plant installation in the *Phytobatea*s and module assembly and attachment to the floating dock, was estimated as 5 min/m^2, done by two untrained workers.

The lifting machine prototype performed well, making *Phytobatea* removal from the waterbody feasible, easy, quick (2 minutes per module) and secure (see Figure 7). *Phytobatea*'s removable character was essential when monitoring and maintenance operations needed to be accomplished, such as for dead plants replacement. Neither the modules nor the plants suffered any damage during manipulation. However, manual handling of the device along the dock was identified as the main feature to be improved.

Figure 6 | View of *Juncus maritimus* saltmarsh after two years, in March 2020, in which significant growth of the surviving plants and aerial biomass stability in *Phytobatea* modules can be seen.

Figure 7 | Tilting machine prototype testing at the beginning of the investigation for *Phytobatea* module removal from the waterbody, and the additional function of the ramp as worktable for plant management and monitoring.

CONCLUSIONS

Results of the first full-scale floating treatment saltmarsh discussed here demonstrate the efficacy of this novel phytotechnology for saline wastewater treatment, as well as its suitability for the implementation of TWs under the harsh and highly variable conditions of contaminated brackish and saline waterbodies. The relatively small size of the FTS (7% of the tidal lagoon total surface), and the low hydraulic retention time (below 1 hour), indicate that the FTS design in the form of a cross-flow barrier is highly efficient, particularly on TP, TSS, BOD, TOC, turbidity and *E. coli* reduction, as well as on wastewater oxygenation. However, it was insufficient for the treatment of nitrogen compounds. Problems with *J. maritimus* survival and growth during the investigation led to an average density of functional plants of 50% of the initial plantation, so it is considered that these results could be significantly improved if the FTS is assessed at full performance, with all plants alive, which would be desirable in the future for a better understanding of these systems' optimal design and dimensioning. In this regard, *J. maritimus* limit for direct exposure to highly saline contaminated water was identified as 38 g/L, and its survival rate for salinities of this magnitude was demonstrated to be dependent on the initial sizes of the perennial parts of the plants. This information increases current knowledge on *J. maritimus* salt tolerance range when planted bare rooted in hydroponic systems, and highlights that, beyond the importance of selecting local, native halophyte plant species for the construction of FTS systems, the size and morphology of the plants could be a critical selection criterion to improve plant survival rates, especially in highly saline environments. This should be considered not only for working with *J. maritimus*, but also with other halophytes. *Phytobatea* technology was validated as an efficient, effective, resistant, long lifespan, and manageable technology for FTS implementation and operation, while *J. maritimus* FTS role as preferential habitat for wild aquatic fauna species revealed the potential of this technology for biodiversity conservation purposes. In addition, there is growing interest from several strategic sectors in *J. maritimus* as raw material for added-value products and services provision, which boosts the feasibility of incorporating *J. maritimus* FTSs within integrated aquaculture systems, hence contributing to aquaculture efficiency, environmental behaviour, and sustainability.

ACKNOWLEDGEMENTS

This research was undertaken with the support given by the Centre for Environmental Research of the Autonomous Community Government of Cantabria (CIMA). The authors would also like to acknowledge CIMA laboratory technicians for their work on the water samples analysis, the partners of Convive LIFE Project for financing the floating treatment saltmarsh design and construction, AC Proyectos and Land Lab studios for their contributions in the design of the construction project, Naútica Parayas for their collaboration with the first prototypes design and manufacture of *Phytobatea* technology, Rubén Caballero for his altruistic contribution producing the artwork for this paper, Eva de la Lama Cicero for her revision of English language, and to the aquaculture company Sonrionansa S.L. for hosting and supporting this project.

DATA AVAILABILITY STATEMENT

All relevant data are included in the paper or its Supplementary Information.

REFERENCES

Alamer, S. F. & Algaraawi, N. I. 2020 Phytochemical profile and antifungal activity of stems and leaves methanol extract from the *Juncus maritimus* Linn. Juncaceae family against some dermatophytes fungi. *AIP Conference Proceedings* **2290**. https://doi.org/10.1063/5.0027554.

Almeida, C. M. R., Mucha, A. P. & Vasconcelos, M. T. S. D. 2006 Comparison of the role of the sea club-rush *Scirpus maritimus* and the sea rush *Juncus maritimus* in terms of concentration, speciation and bioaccumulation of metals in the estuarine sediment. *Environmental Pollution* **142** (1), 151–159. https://doi.org/10.1016/j.envpol.2005.09.002.

Boscaiu, M., Ballesteros, G., Naranjo, A. & Vicente, O. 2011 Responses to salt stress in *Juncus acutus* and *J. maritimus* during seed germination and vegetative plant growth. *Plant Biosystems* **145** (4), 770–777.

Boyd, C. E., D'Abramo, L. R., Glencross, B. D., Huyben, D. C., Juarez, L. M., Lockwood, G. S., McNevin, A. A., Tacon, A. G. J., Teletchea, F., Tomasso, J. R., Tucket, C. S. & Valenti, W. C. 2020 Achieving sustainable aquaculture: historical and current perspectives and future needs and challenges. *Journal of the World Acuaculture Society* **51**, 578–633. https://doi.org/10.1111/jwas.12714.

Calheiros, C. S. C., Carecho, J., Tomasino, M. P., Almeida, C. M. R. & Mucha, A. P. 2020 Floating wetland islands implementation and biodiversity assessment in a Port Marina. *Water* **12** (11), 3273. https://doi.org/10.3390/w12113273.

Castillo, J., Rubio-Casal, A. E. & Figueroa, E. 2010 Cordgrass biomass in salt marshes. In: *Biomass* (Momba, M. N. B., ed.). IntechOpen. https://doi.org/10.5772/9767.

Chartzoulakis, K. & Klapaki, G. 2000 Response of two greenhouse pepper hybrids to NaCl salinity during different growth stages. *Scientia Horticulturae* **86**, 247–260. https://doi.org/10.1016/S0304-4238(00)00151-5.

Cicero-Fernandez, D., Expósito-Camargo, J. A., Peña-Fernandez, M. & Antizar-Ladislao, B. 2019 *Carex paniculata* constructed wetland efficacy for stormwater, sewage and livestock wastewater treatment in rural settlements of mountain areas. *Water Science Technology* **79** (7), 1338–1347. https://doi.org/10.2166/wst.2019.130.

Colares, G., Dell'Osbel, N., Wiesel, P., Oliveira, G., Lemos, P., Silva, F., Lutterbeck, C., Kist, L. & Machado, Ê. 2020 Floating treatment wetlands: a review and bibliometric analysis. *Science of The Total Environment* **714**, 136776. https://doi.org/10.1016/j.scitotenv.2020.136776.

De Lange, M., Paulissen, M. & Slim, P. 2013 'Halophyte filters': the potential of constructed wetlands for application in saline aquaculture. *International Journal of Phytoremediation* **15**, 352–364. doi:10.1080/15226514.2012.702804.

Delattre, E., Techer, I. & Reneaud, B. 2021 Chloride accumulation in aboveground biomass of three macrophytes (*Phragmites australis*, *Juncus maritimus*, and *Typha latifolia*) depending on their growth stages and salinity exposure. Application for Cl- removal and phytodesalinization. *Environmental Science and Pollution Research*. (In Review). https://doi.org/10.21203/rs.3.rs-158199/v1

Doody, J. 2010 *Management of Natura 2000 Habitats Atlantic Salt Meadows (Glauco-Puccinellietalia maritimae) 1330: Directive 92/43/EEC on the Conservation of Natural Habitats and of Wild Fauna and Flora*. European Commission. https://data.europa.eu/doi/10.2779/50003

Fernandez, E. & Marquínez, J. 2002 Zonación morfodinámica e incidencia antrópica en los estuarios de Tina Mayor y Tina Menor (Morphodynamic zoning and anthropic incidence in Tina Mayor and Tina Menor Estuaries). *Revista de la Sociedad Geológica de España* **15**, 141–157.

Hassan, M. A., López-Gresa, M. P., Boscaiu, M. & Vicente, O. 2016 Stress tolerance mechanisms in *Juncus*: responses to salinity and drought in three *Juncus* species adapted to different natural environments. *Functional Plant Biology* **43**, 949–960.

Headley, T. & Tanner, C. 2008 Floating treatment wetlands: an innovative option for stormwater quality applications. In: *11th International Conference on Wetland Systems for Water Pollution Control*, pp. 1101–1106.

Headley, T. & Tondera, K. 2019 Floating treatment wetlands. In: *Wetland Technology. Practical Information on the Design and Application of Treatment Wetlands* (Langergraber, G., Dotro, G., Nivala, J., Rizzo, A., & Stein, O. R., eds) IWA Publishing, London. Scientific and Technical Report, 27, pp. 112–114.

Huang, X., Zhao, F., Song, C., Chai, Y., Wang, Q. & Zhuang, P. 2020 Larva fish assemblage structure in three-dimensional floating wetlands and non-floating wetlands in the Changjiang River estuary. *Journal of Oceanology and Limnology* **39**. doi:10.1007/s00343-020-0078-6.

Idrisi, A. H., Mourad, A. H. I. & Sherif, M. M. 2021 Impact of prolonged exposure of eleven years to hot seawater on the degradation of a thermoset composite. *Polymers* **13**, 2154. https://doi.org/10.3390/polym13132154.

Karstens, S., Langer, M., Nyunoya, H., Čaraitė, I., Stybel, N., Razinkovas-Baziukas, A. & Bochert, R. 2021 Constructed floating wetlands made of natural materials as habitats in eutrophicated coastal lagoons in the Southern Baltic Sea. *Journal of Coastal Conservation* **25** (44). https://doi.org/10.1007/s11852-021-00826-3.

Kourkoumpas, D. S., Curt, M., Kallen, S., Buevin, M., Zapatero, M., Mardikis, M., Kiartzis, S., Díaz, P., Aguado, P. L., Bezergianni, S. & Panagiotis, G. 2019 Assessment of the potential of green floating filters for bioenergy production. In: *Proceedings of the 26th European Biomass Conference and Exhibition*, pp. 209–213.

Kristensen, P., Kampa, E. & Völker, J. 2021 *Drivers of and Pressures Arising From Selected Key Water Management Challenges: A European Overview*. European Environment Agency. Report No 9/2021. https://data.europa.eu/doi/10.2800/059069.

Liang, Y., Zhu, H., Bañuelos, G., Yan, B., Zho, Q., Yu, X. & Cheng, X. 2017 Constructed wetlands for saline wastewater treatment: a review. *Ecological Engineering* **98**, 275–285. https://doi.org/10.1016/j.ecoleng.2016.11.005.

Lissner, J. & Schierup, H.-H. 1997 Effects of salinity on the growth of *Phragmites australis*. *Aquatic Botany* **55**, 247–260. https://doi.org/10.1016/S0304-3770(96)01085-6.

López-Álvarez, M., Pereiro, I., Serra, J., González, P. & de Carlos-Villamarín, A. 2012 Porous silicon carbide scaffolds with patterned surfaces obtained from the Sea Rush *Juncus maritimus* for tissue engineering applications. *International Journal of Applied Ceramic Technology* **9** (10), 1744–7402.

Lotze, H., Lenihan, H., Bourque, B., Bradbury, R., Cooke, R., Kay, M., Kidwell, S., Kirby, M., Peterson, C. & Jackson, J. 2006 Depletion, degradation, and recovery potential of estuaries and coastal seas. *Science* **312** (5781), 1806–1809. https://www.science.org/doi/abs/10.1126/science.1128035.

Marques, B., Lillebø, A. I., Pereira, E. & Duarte, A. C. 2011 Mercury cycling and sequestration in salt marshes sediments: an ecosystem service provided by *Juncus maritimus* and *Scirpus maritimus*. *Environmental Pollution* **159** (7), 1869–1876. https://doi.org/10.1016/j.envpol.2011.03.036.

Montenegro, I. P. F. M., Mucha, A. P., Reis, I., Rodrigues, P. & Almeida, C. M. R. 2016 Effect of petroleum hydrocarbons in copper phytoremediation by a salt marsh plant (*Juncus maritimus*) and the role of autochthonous bioaugmentation. *Environmental Science Pollution Restoration* **23**, 19471–19480. https://doi.org/10.1007/s11356-016-7154-7.

Pavlineri, N., Skoulikidis, N. T. & Tsihrintzis, V. A. 2017 Constructed floating wetlands: a review of research, design, operation and management aspects, and data meta-analysis. *Chemical Engineering Journal* **308**, 1120–1132.

Radermacher, M., Wit, L., Winterwerp, J. & Uijttewaal, W. 2015 Efficiency of hanging silt curtains in crossflow. *Journal of Waterway, Port, Coastal, and Ocean Engineering* **142**, 04015008. doi:10.1061/(ASCE)WW.1943-5460.0000315.

Ribeiro, H., Almeida, C. M. R., Mucha, A. P., Teixeira, C. & Bordalo, A. A. 2013 Influence of natural rhizosediments characteristics on hydrocarbons degradation potential of microorganisms associated to *Juncus maritimus* roots. *International Biodeterioration & Biodegradation* **84**, 86–96. https://doi.org/10.1016/j.ibiod.2012.05.039.

Ruiz, J. M., León, V. M., Marín, L., Giménez, F., Rogel, J. A., Esteve, M. A., Gómez, R., Robledano, F., González, G. & Martínez, J. 2020 Informe de síntesis sobre el estado actual del Mar Menor y sus causas en relación a los contenidos de nutrientes (Synthesis report on the current state of the Mar Menor and its causes in relation with nutrient content). *Juezas y jueces para la Democracia, Boletín comisión contencioso administrativo* **3** (2), 4–12.

Saghrouni, Z., Baillis, D. & Jemni, A. 2020 Composites based on *Juncus maritimus* fibers for building insulation. *Cement and Concrete Composites* **106**, 103474. ISSN 0958-9465. https://doi.org/10.1016/j.cemconcomp.2019.103474.

Sahli, R., Rivière, C. & Siah, A. 2018 Biocontrol activity of effusol from the extremophile plant, *Juncus maritimus*, against the wheat pathogen *Zymoseptoria tritici*. *Environmental Science and Pollution Research* **25**, 29775–29783. https://doi.org/10.1007/s11356-017-9043-0.

Sahuc, M.-E., Sahli, R., Rivière, C., Pène, V., Lavie, M., Vandeputte, A., Brodin, P., Rosenberg, A. R., Dubuisson, J., Ksouri, R., Rouillé, Y., Sahpaz, S. & Séron, K. 2019 Dehydrojuncusol, a natural phenanthrene compound extracted from *Juncus maritimus*, is a new inhibitor of hepatitis C virus RNA replication. *Journal of Virology* **93** (10), e02009-18. https://doi.org/10.1128/JVI.02009-18.

Sanicola, O., Lucke, T., Stewart, M., Tondera, K. & Walker, C. 2019 Root and shoot biomass growth of constructed floating wetlands plants in saline environments. *International Journal of Environmental Research and Public Health* **16** (2), 275. https://doi.org/10.3390/ijerph16020275.

Schwammberger, P., Walker, C. & Lucke, T. 2017 Using floating wetland treatment systems to reduce stormwater pollution from urban developments. *International Journal of GEOMATE* **12**, 45–50.

Shahid, M., Muhammad, A., Shafaqat, A., Siddique, M. & Muhammad, A. 2018 Floating wetlands: a sustainable tool for wastewater treatment. *Clean – Soil Air Water*. doi:10.1002/clen.201800120.

Sharma, R., Vymazal, J. & Malaviya, P. 2021 Application of floating treatment wetlands for stormwater runoff: a critical review of the recent developments with emphasis on heavy metals and nutrient removal. *Science of The Total Environment* **777**, 146044. doi:10.1016/j.scitotenv.2021.146044.

Shen, S., Li, X. & Lu, X. 2021 Recent developments and applications of floating treatment wetlands for treating different source waters: a review. *Environmental Science and Pollution Research* **28**. doi:10.1007/s11356-021-16663-8.

Smichi, N., Messaoudi, Y., Moujahed, N. & Gargouri, M. 2015 Ethanol production from halophyte *Juncus maritimus* using freezing and thawing biomass pretreatment. *Renewable Energy* **85**. doi:10.1016/j.renene.2015.07.010.

Syranidou, E., Christofilopoulos, S. & Kalogerakis, N. 2016 *Juncus* spp.–The helophyte for all (phyto)remediation purposes?. *New Biotechnology* **38**, 43–55. https://doi.org/10.1016/j.nbt.2016.12.005.

Tanner, C. C. & Headley, T. R. 2011 Components of floating emergent macrophyte treatment wetlands influencing removal of stormwater pollutants. *Ecological Engineering* **37** (3), 474–486.

Vymazal, J. 2007 Removal of nutrients in various types of constructed wetlands. *Science of the Total Environment* **380**, 48–65.

Vymazal, J., Zhao, Y. & Mander, Ü. 2021 Recent research challenges in constructed wetlands for wastewater treatment: a review. *Ecological Engineering* **169**, 106318. https://doi.org/10.1016/j.ecoleng.2021.106318.

Ware, J. & Callaway, R. 2019 Public perception of coastal habitat loss and habitat creation using artificial floating islands in the UK. *PLoS ONE* **14**, 1–16. doi:10.1371/journal.pone.0224424.

Yahiaoui, K., Ouakouak, A., Guerrouf, N., Zoubeidi, A. & Hamdi, N. 2020 Domestic wastewater treatment by vertical-flow filter grown with *Juncus Maritimus* in Arid Region. *International Journal of Engineering Research in Africa* **47**, 109–117. https://doi.org/10.4028/www.scientific.net/jera.47.109.

Yajun, C., Ya, Z., Naiwei, L., Xiaojing, L., Fengfeng, D. & Dongrui, Y. 2019 Biomass production and nutrient removal efficiency of *Suaeda salsa* in eutrophic saline water using a floating mat treatment system. *Water Supply* **19** (1), 254–263. https://doi.org/10.2166/ws.2018.066.

Yang, L. 2019 Saline Treatment Wetlands. In: *Wetland Technology. Practical Information on the Design and Application of Treatment Wetlands* (Langergraber, G., Dotro, G., Nivala, J., Rizzo, A., & Stein, O. R., eds). IWA Publishing, London. Scientific and Technical Report No. 27, pp. 70–72.

Zal, N., Whalley, C., Christiansen, T., Kristensen, P. & Néry, F. 2018 *European Waters. Assessment of Status and Pressures 2018*. European Environment Agency. Report No 7/2018. doi:10.2800/303664.

Zhang, L., Zhao, J. & Cui, N. 2016 Enhancing the water purification efficiency of a floating treatment wetland using a biofilm carrier. *Environmental Science Pollution Research* **23**, 7437–7443.

First received 4 January 2022; accepted in revised form 25 April 2022. Available online 5 May 2022

doi: 10.2166/wst.2022.099

Toxicity and removal of pharmaceutical and personal care products: a laboratory scale study with tropical plants for treatment wetlands

Alexander Arredondo[a], Carlos A. Ramírez-Vargas [b,c,*], Janneth A. Cubillos[a], Juan P. Arrubla[d], Tito Morales-Pinzón[e], Diego Paredes[a] and Carlos A. Arias[b,c]

[a] Water and Sanitation Research Group, Faculty of Environmental Sciences, Universidad Tecnológica de Pereira, Pereira A.A. 97, Colombia
[b] Department of Biology, Aarhus University, Ole Worms Allé 1, Building 1135, Aarhus C. 8000, Denmark
[c] WATEC, Aarhus University, Ny Munkegade 120, Aarhus C 8000, Denmark
[d] Water Resources Study Group, School of Chemistry, Universidad Tecnológica de Pereira, Pereira A.A. 97, Colombia
[e] Territorial Environmental Management Research Group, Faculty of Environment Sciences, Universidad Tecnológica de Pereira, Pereira A.A. 97, Colombia
*Corresponding author. E-mail: c.a.ramirez@bio.au.dk

CAR-V, 0000-0003-0614-9490

ABSTRACT

The aim of the research was to evaluate the response of three tropical species (*Heliconia psittacorum, Ciperus haspan, Hedychium coronarium*), respect their tolerance and removal capacity of pharmaceutical and personal care products (PPCPs), namely acetylsalicylic acid, ibuprofen, methyl hydrojasmonate (cis – MDJM), galaxolide, tonalide, caffeine, naproxen, ketoprofen, and diclofenac. The study was undertaken in two stages (Stage I – Tolerance; Stage II – Removal) of 21 days each. In Stage I, it was found evidence that from $1,000\ \mu g\ L^{-1}$ the plants show decaying responses, being *C. haspan* and *H. psittacorum*, the species with the best responses to tolerance and adaptation. The results of Stage II indicated that tonalide and ketoprofen compounds were 99% removed during the first 24 hours of exposure; acetylsalicylic acid, ibuprofen, galaxolide, and naproxen compounds were 80% eliminated, and caffeine and diclofenac products presented lower removal rates during same time. The study allowed the identification of two compound blocks, PPCPs that are sorbed by plants (acetylsalicylic acid, ibuprofen, MDJM, caffeine, galaxolide, and tonalide), and highly photodegradable compounds (ketoprofen, naproxen, and diclofenac). These findings open the possibility for further research about using plants adapted to tropical conditions, for PPCP removal from wastewaters in real scale nature-based systems such as treatment wetlands.

Key words: emerging pollutants, microcontaminants, physiological responses, plant species, removal mechanisms, tolerance

HIGHLIGHTS

- The toxicity tolerance and removal of PPCPs were tested with 3 tropical plants typically used in treatment wetlands. Max. tolerance was determined between 100 and $1,000\ \mu g\ L^{-1}$.
- The highest PPCPs removal occurred during the first 24 hours of exposure.
- The presence of macrophytes in assessed reactors favors the removal of PPCPs and has the potential for being implemented in real-scale wetland systems in tropical areas.

GRAPHICAL ABSTRACT

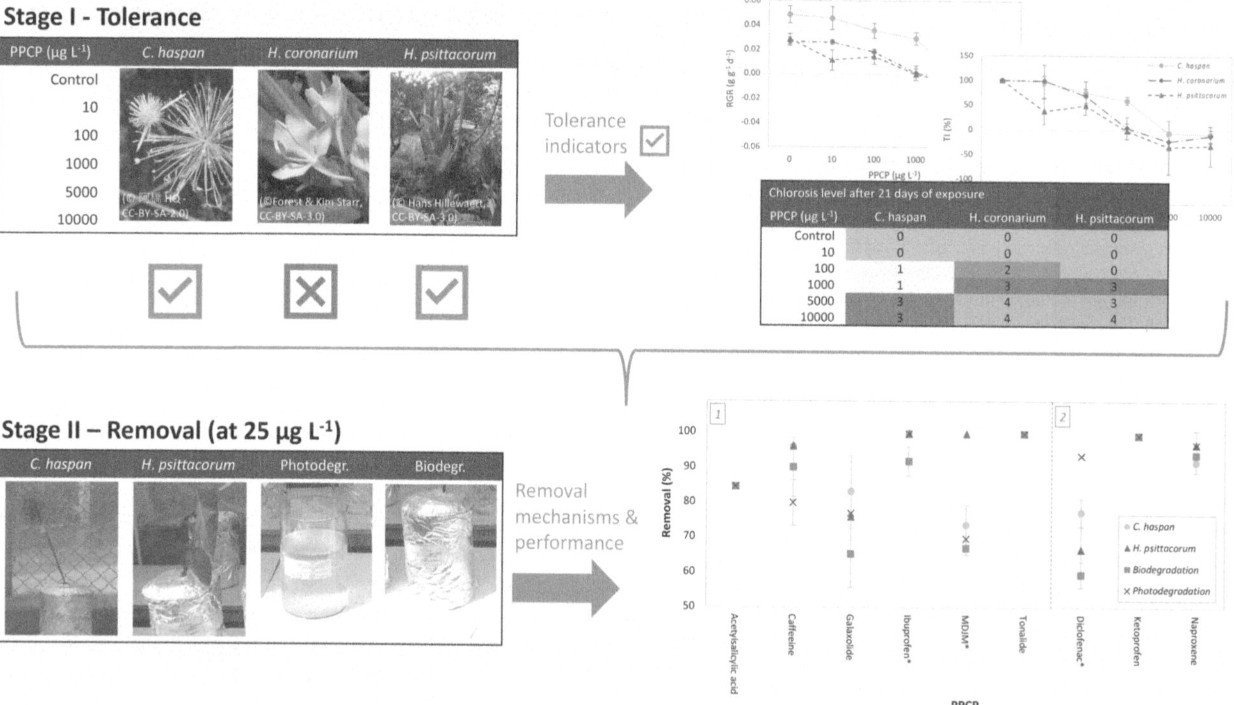

1. INTRODUCTION

Pharmaceutical and personal care products (PPCPs) are chemically and biologically active substances that can be persistent in the environment (Santos *et al.* 2010). They are a part of group of substances known as emerging contaminants. The United States Environmental Protection Agency (USEPA) defines them as new chemical products without regulation that generate an unknown impact in their surroundings (Deblonde *et al.* 2011). Concerns about the presence of PPCPs and their possible effects in aquatic ecosystems have arisen in the past years, since continuous discharge of these compounds may cause accumulation and irreversible damage to wildlife and human health (Liu & Wong 2013; Ortiz *et al.* 2013); however, the effects on human beings are not well identified to date (Santos *et al.* 2010).

PPCPs identified in natural environments belong to a diverse group of organic substances. Sources of these are pharmaceutical products such as analgesics, antibiotics, anticonvulsants, lipid regulators, and hormones, and daily use personal care products such as soaps, lotions, sun blocking lotions, and toothpaste, among others (Liu & Wong 2013). The main source of these compounds to nature is through inadequate disposal of unused or expired medicines from homes and hospitals, by excretion, indiscriminate use of fertilizers, fungicides and pesticides in agriculture, and direct discharge of veterinary products (Santos *et al.* 2010). In those ways, they reach the aquatic environments through point source discharges such as effluents from sewer system and wastewater treatment plants (Zhang *et al.* 2014), as well as from non-point sources such as rainfall runoff and leachates from agricultural and livestock production (Mompelat *et al.* 2009).

Conventional technologies for wastewater treatment and drinking water treatment for human consumption cannot efficiently remove this type of compound (Dordio *et al.* 2010), so these substances remain in the water, even after wastewaters have been treated, or surface waters processed to drinking water (USEPA 2002). More advanced treatment technologies such as ozonation, advanced oxidation, activated carbon, nanofiltration, and reverse osmosis can remove most of the PPCPs present in wastewaters reaching elimination rates above 90% (Sui *et al.* 2010; Yang *et al.* 2011; WHO 2012). However, these technologies are not widely used due to their high cost and operational and maintenance demands (Bolong *et al.* 2009).

Treatment wetlands are systems widely used, of relative low cost for secondary and tertiary treatment of wastewaters (Matamoros *et al.* 2008; Dordio *et al.* 2010; Li *et al.* 2014), and having shown successful results in the elimination of PPCPs (Matamoros *et al.* 2007, 2009; Ávila *et al.* 2013; Verlicchi *et al.* 2013). In this type of systems, the interactions of plants, microorganisms, and water, allow the removal of contaminants from wastewaters, ranging from conventional pollutants to persistent compounds (Zhang *et al.* 2015). Macrophytes carry out chemical, physical and biological processes such as volatilization, absorption, accumulation, adsorption, translocation, and degradation for the elimination of the pharmaceutical compounds (Zhang *et al.* 2014).

Dordio *et al.* (2010) reported that plant species such as *Thypa sp.* in hydroponic environments reach removal efficiencies of 96, 97, and 75% for ibuprofen, carbamazepine, and clofibric acid, respectively. Zhang *et al.* (2013a) reported absorption as the principal elimination mechanism for caffeine in hydroponic environments planted with *Schoenoplectus tabernaemontani*. On the other hand, the results from Matamoros *et al.* (2008) indicate that ibuprofen degrades 51% in subsurface flow treatment wetlands, highlighting that low aerobic processes in the systems evaluated were the major contributors to ibuprofen degradation. Studies performed by Hijosa-Valsero *et al.* (2010) ascribe the elimination of naproxen, ibuprofen, diclofenac, carbamazepine, methyl dihydrojasmonate, galaxolide, and tonalide compounds to the presence of plants in treatment wetlands because they have the capacity to retain particles in suspension, release O_2, serve as support for biofilm, and insulate during low temperatures. For Zhang *et al.* (2014) the principal removal mechanisms for PPCPs are photolytic degradation, sorption over sludge or particles, filtration, chemical oxidation, photocatalysis, and microbial degradation. In treatment systems based on plants, the absorption and the translocation are the primary mechanism for removal.

The diffusion processes of PPCPs in plant tissues depend on their physical–chemical characteristics, which include the compounds' hydrophobicity (K_{WO}) and water solubility. Molecules with K_{WO} values between 0.5 and 3.5 present moderate hydrophobicity allowing the compound to move freely in the plant between the lipid layer and the aqueous phase. In contrast, highly hydrophobic compounds ($K_{WO}>3.5$) are more prone to be absorbed in the roots of plants, while highly polar drugs with values of $K_{WO}<0.5$ are incorporated by the plant and degraded by different metabolic processes (phytodegradation) (Dordio & Carvalho 2013; Li *et al.* 2014; Zhang *et al.* 2014). Most recently, Brunhoferova *et al.* (2021) tested the removal of a mixture of 27 micropollutants (pharmaceuticals, pesticides, herbicides, fungicides, and others) from aqueous solution by phytoremediation, with comparable and even higher removal efficiency than conventional wastewater treatment plants. The main objective of this study was to assess the tolerance and elimination capacity of PPCPs in planted microcosm reactors with *Heliconia psittacorum, Ciperus haspan, and Hedychium coronarium*.

2. MATERIALS AND METHODS

2.1. Tested compounds

The PPCPs selected for the study were: acetylsalicylic acid (aspirin), ibuprofen, naproxen, diclofenac, ketoprofen, caffeine, diclofenac, methyl-hydrojasmonate (cis – MDJM), galaxolide, and tonalide, which are products frequently consumed and freely available in the market. The products were commercially acquired in local pharmacies in Colombia, which distribute medicines manufactured by Genfar, MK, La Santé, and Bayer. Table 1 presents the characteristics of each of the evaluated compounds.

2.2. Experimental units

The study was carried out in two stages: Stage I – Tolerance, and Stage II – Removal. For Stage I, the species selected were *H. psittacorum, C. haspan* and *H. coronarium*, common wetlands plants in tropical areas. The plants were selected and collected from natural wetlands, and transported to laboratory, where their roots were thoroughly washed with distilled water to remove soil and humic compounds. Eighteen individuals of similar size of the studied species were randomly selected to be hydroponically planted in 1 liter glass reactors lined with aluminium foil (to inhibit algae development).

The reactors were filled with an adapted Hoagland nutrient solution (Alkhatib *et al.* 2013) to guarantee the nutritional requirements of the plants (202 g L^{-1} KNO$_3$, 147 g L^{-1} CaCl$_2$, 15 g L^{-1} Fe-EDTA, 493 g L^{-1} MgSO$_4$.7H$_2$O, 159.98 g L^{-1} NH$_4$NO$_3$, 2.86 g L^{-1} H$_3$BO$_3$, 1.81 g L^{-1} MnCl.4H$_2$O, 0.22 g L^{-1} ZnSO$_4$.7H$_2$O, 0.051 g L^{-1} CuSO$_4$, 0.12 g L^{-1} Na$_2$MoO$_4$, and 136 g L^{-1} KH$_2$PO$_4$). After an adaptation period of 3 weeks in hydroponic conditions, six groups (triplicates) of reactors per each plant species were stablished. Five groups of reactors were spiked with a PPCP cocktail of the selected compounds at different concentrations (10, 100, 1,000, 5,000, and 10,000 by PPCP); likewise, one group of blanks (triplicate) filled only with nutrient solution (0 µg/L by PPCP) were placed as controls. The reactors were exposed to natural environmental conditions (under translucid shelter to protect against rainfall and allow the passage of natural light), with day/night cycles of ± 12 h, and temperature ranging from 24 °C (day) to 17 °C (night). A total of 54 reactors were monitored (Table 2).

Table 1 | Characteristics of studied PPCPs

Compound	Category	Chemical structure	CAS number	Physicochemical properties		
				PM	S_{H2O}	Log K_{ow}
Acetylsalicylic Acid (Aspirin)	Analgesic/Anti inflammatory		69-72-7	138.12	2240	2.26
Ibuprofen	Anti inflammatory		15687-27-1	206.29	21	3.97
Naproxen	Anti inflammatory		22204-53-1	230.27	15.9	3.18
Diclofenac	Anti inflammatory		15307-86-5	296.16	2.37	4.57
Ketoprofen	Anti inflammatory		22071-15-4	254.29	51	3.12
Caffeine	Stimulant		58-08-2	194.19	2.16*104	-0.07
Methyl–hydrojasmonate	Fragrance ingredient		24851-98-7	226.31	280	2.98
Galaxolide	Fragrance ingredient		1222-05-05	258.4	1.75	5.9
Tonalide	Fragrance ingredient		21145-77-7	258.41	1.25	5.7

Source: Adapted from Matamoros *et al.* 2012.

The reactors corresponding to Stage II were set up with the two species showing the best tolerance in Stage I. From each species two similar-sized individuals were taken and placed in 8-liter reactors, lined with aluminum foil and fed with Hoagland nutrient solution (Alkhatib *et al.* 2013). The reactors were spiked with the same PPCP cocktail used in Stage I, but with a concentration of 25 µg/L by compound. The photodegradation effect was determined by setting up two blank reactors without plants and exposed to the same environmental conditions of the rest of the mesocosms. Two additional reactors without plants and covered were set up to determine the microbial degradation effect; 8 reactors in total were

Table 2 | Experimental set-up for tolerance test (stage I) and removal capacity test (stage II) of selected tropical plants against different PPCPs concentrations

Stage	Factors	Levels	Number of levels	Replicates	Reactors
I	Concentration (µg L^{-1})	0 10 100 1,000 5,000 10,000	6	3	54
	Plants	*C. haspan* *H. coronarium* *H. psittacorum*	3		
II	Removal mechanism (at 25 µg L^{-1})	Phytodegradation – Plant 1 Phytodegradation – Plant 2 Biodegradation Photodegradation	4	2	8

implemented to monitor the PPCPs' elimination (Table 2). The reactors were submitted to the same environmental conditions as in Stage I.

2.3. Exposure period and measurement strategy

The total period of experiment was 42 days. Stage I was carried out during 21 days, and included the measurements of stem length, root length, and biomass on days 0, 3, 7, 14, and 21, to estimate the relative growth rate (RGR; Equation (1)), the tolerance index (TI; Equation (2)), the growth effect (GE; Equation (3)), and chlorosis (visual comparison).

$$RGR \ (g \ g^{-1} \ d^{-1}) = \ (\ln W_2 - \ln W_1)/T*100 \tag{1}$$

$$TI \ (\%) = RGR_c/RGR_0 * 100 \tag{2}$$

$$GE \ (\%) = L_c/L_0 * 100 \tag{3}$$

where, W_1 is the wet weight (g) of the plant at the beginning of the experiment, W_2 is the wet weight (g) of the plant at the end of the experiment, and T is the test time (days); RGR_c is the relative growth rate (g g^{-1} d^{-1}) in reactors contaminated with PPCP, and RGR_0 is the relative growth rate (g g^{-1} d^{-1}) in control reactors; L_c is the length (cm) of roots or stems in reactors contaminated with PPCP, and L_0 is the length (cm) of roots and stems in control reactors.

Stage II also had an experimental period of 21 days, where grab water samples were taken on day 1, 3, 12 and 21. For the sampling, each unit was homogenized, and 1,000 mL of sample were taken and bottled in amber color glass bottles to be preserved at 4 °C (Standard Methods – APHA 2012). Thereafter, the samples were analyzed by the Laboratory of Chromatography and Mass Spectrometry – CROMMAS of Universidad Industrial de Santander – UIS (Colombia).

2.4. Solid phase extraction (SPE)

The water samples were filtered through fiberglass filters with a pore diameter size of 0.45 µm and 47 mm, and subsequently acidified with concentrated chloride acid up to pH<2. Afterward, 500 mL were percolated with polymeric PE cartridges packed with 100 mg of Strata X, previously prepared with 5 mL of n-hexane, 5 mL of ethyl acetate, 10 mL of MeOH and 10 mL of Milli Q water. The filtration rate was adjusted to approximately 10 mL min^{-1}. Then, the cartridges were dried and finally the analytes were eluted with 5 mL of ethyl acetate. The extract was evaporated until dried completely under a soft flow of reconstituted nitrogen in 175 µL of methanol (Matamoros & Bayona 2006).

2.5. Chromatographic analysis

The sample extracted in the SPE phase was injected into an AT 6890 series plus chromatographer (Agilent Technologies, Palo alto California, USA) coupled with a selective mass detector (Agilent Technologies, MSD 5797 Inert XL) operated in SIM mode (selected ion monitoring). The column employed in the analysis was DB-5MS ((5% phenyl)-methylpolysiloxane,

60 m$\times$0.25 mm$\times$0.25 µm). The injection was carried out in the splitless mode (Vol inj.$=$2 µL). The column used was capillary HP-5 (5% phenyl, 95% dimethyl-polysiloxane) (30 m$\times$0.25 mm ID, 0.25 µm thickness phase), the He gas travel was 1.0 mL min^{-1}, and the injection volume was 2 µL in splitless mode. The methylation of the carboxylate acid group was carried out in line in a hot gas injector with an added 10 µL solution of trimethylsulfonium hydroxide TMSH (0.25 mol L^{-1} in methanol) to 50 µL of sample before the injection. In the ion selective monitoring mode the following ions were selected:

Methyl ester salicylic acid (92/120/121/152), methyl ester ibuprofen (161/162/177/220), methyl ester naproxen (170/185/186/244), methyl ester diclofenac (214/216/242/309), methyl ester ketoprofen (191/209/210/268), caffeine (82/109/165/194), dihydrojasmonate (83/153/156/226), galaxolide (213/243/244/258), tonalide (201/243/244/258).

2.6. Data analysis

To evaluate the results of the impact of PPCP concentrations over biomass growth of the tested plant species, a non-parametric analysis was carried out using the Mann Whitney test. ($P<0.05$). An ANOVA and Tukey HSD analysis was implemented ($P<0.05$) to identify significant differences in the removal of PPCP among phytodegradation, biodegradation, and photodegradation reactors. The results were analyzed with the statistical package XLSTAT 2021.5.1.1204 (Addinsoft 2021).

3. RESULTS AND DISCUSSION

3.1. Stage I – tolerance

Figure 1 presents results of RGR of tested plants exposed to different PPCP concentrations. After 21 days of test, the results showed a decreasing trend in biomass gain for all the plants as the spiked PPCP concentration increases. At 1,000 µg L^{-1} *H. psittacorum* and *H. coronarium* plants showed a strong growth inhibition in comparison to control plants (at 0 µg L^{-1}), with a depletion in RGR of 101 and 97% respectively (without significant differences among them), whereas *Ciperus haspan* was the least affected plant with a decrease in RGR of $\pm$40%.

When reactors were spiked with PPCPs from 5,000 µg L^{-1}, all the tested plants showed a decrease in RGR above 100%, denoting a loss of vegetal biomass. Likewise, the tolerance index (TI%) showed that from 1,000 µg L^{-1}, *H. psittacorum* and *H. coronarium* plants presented marked symptoms of toxicity, while for *C. haspan* plants, the same marked toxicity signs were expressed from PPCP concentration of 5,000 µg L^{-1} (Figure 1).

The impact of different PPCP concentrations on the growth of stems and roots (growth effect – GE) can be seen in Figure 2. Similarly to the impact of different PPCPs concentration on RGR, the GE of *H. psittacorum* and *H. coronarium* plants was affected from concentrations of 1,000 µg L^{-1}, a fact evident after the second monitoring week, when growth inhibition was observed and represented in biomass losses. After 21 days of experiment, the average GE of stems for *H. psittacorum* and *H. coronarium* spiked with PPCPs at 1,000 µg L^{-1} were $\pm$85% and $\pm$99% lower than in control plants (0 µg L^{-1}), whereas the average GE of stems *C. haspan* was $\pm$25% higher than in control plants. From 5,000 µg L^{-1} all tested species showed stem growth inhibition and loss of height due to withering. Regarding the GE of roots, in *C. haspan* it was $\pm$91% lower that in

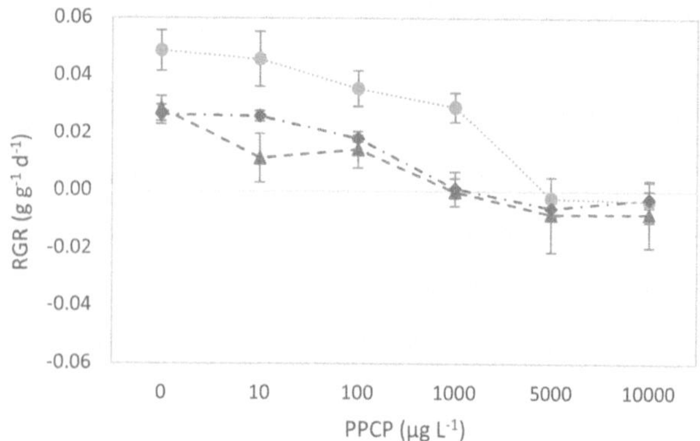
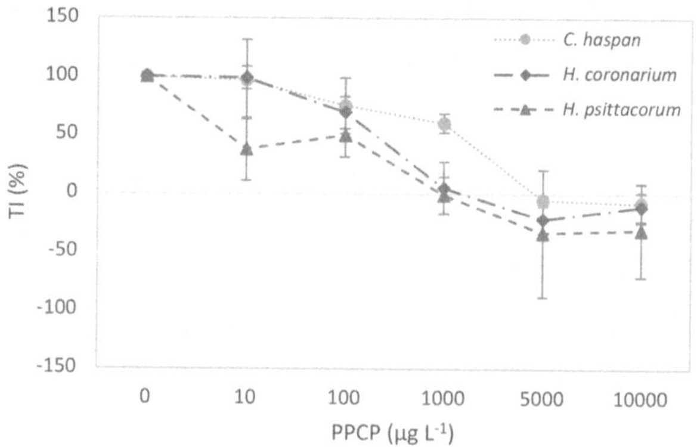

Figure 1 | Relative growth rate (RGR) and tolerance index (TI) by plant species and PPCP concentration after 21 days of experimental Stage I. Bars indicate standard deviation.

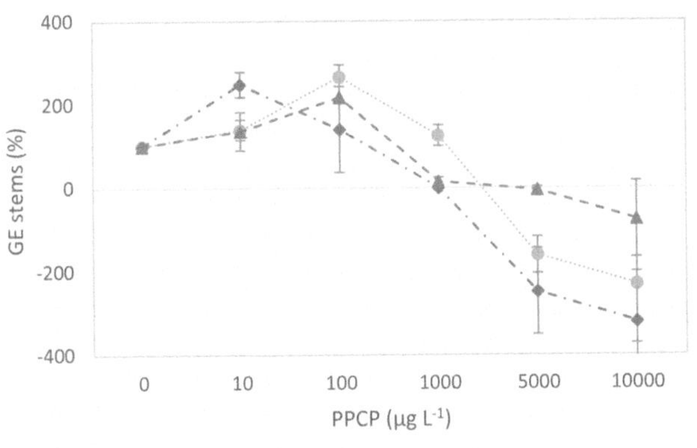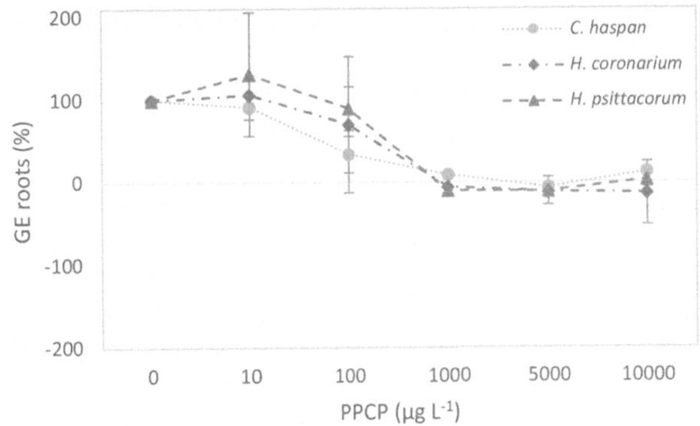

Figure 2 | Growth effect (GE) of stems and roots by plant species and PPCP concentration after 21 days of experimental Stage I. Bars indicate standard deviation.

control plants (0 µg L^{-1} PPCPs), whereas the GE of roots for *H. psittacorum* and *H. coronarium* showed growth inhibition and loose in length due to withering when spike with PPCPs at 1,000 µg L^{-1}.

Table 3 presents a comparison of the appearance of chlorosis among tested plant species. The results indicated that after 21 days of exposure, in reactors spiked with PPCPs concentrations below 100 µg L^{-1}, the plants did not show evidence of chlorosis. At 100 µg L^{-1} concentration, *C. haspan* and *H. coronarium* plants showed 'low' and 'middle' signs of chlorosis respectively, signs that start to be evident with 'low' chlorosis levels since day 7; after 21 days of experiment *H. psittacorum* did not show chlorosis signs when spiked with PPCPs at 100 µg L^{-1}. At the end of the test the *H. coronarium* and *H. psittacorum* plants displayed 'high' chlorosis symptoms when spiked with PPCP concentrations above 1,000 µg L^{-1}, whereas *C. haspan* plants just reach a 'high' chlorosis level. For PPCP concentration of 10,000 µg L^{-1}, *H. coronarium* and *H. psittacorum* plants reached signs of 'necrosis'. Control plants fed only with Hoagland nutrient solution did not show any symptom of chlorosis.

Regarding growth expressed as biomass, there are statistically significant effects ($p<0.05$) in *C. haspan* and *H. coronarium* (Table 4), while the effect of the concentration over biomass growth presents significant differences on a $p<0.05$ between concentrations of 0 and 1,000 µg/L, and the corresponding concentrations to 5,000 and 10,000 µg/L (Table 5). These differences are corroborated through a hierarchical conglomerate analysis that allowed identifying and classifying two main groups in agreement to their concentrations (Figure 3).

The loss of biomass in the plants exposed to high PPCP concentrations may be explained by the decrease of the redox capacity on the cellular membranes that are affected by the exposition to xenobiotic agents and cause the increase of reactive oxygen species that drive the cellular death of the plant (Iori *et al.* 2012). Another physiological response to the stress caused by the added pollutants is reflected by the inhibition of radicular growth that disturbs the plant processes making it prone to diseases and inducing changes in the chlorophyll content, thus decreasing its photosynthetic capacity (Nadeem *et al.* 2014).

The results of Stage I allow the conclusion that the species *C. haspan and H. psittacorum* and are the most tolerant to exposure to high PPCP concentration, and are therefore fit to be planted in treatment wetlands for the elimination of PPCPs.

Table 3 | Chlorosis responses to different concentrations of PPCPs over the species *C. haspan*, *H. coronarium*, and *H. psittacorum*, during 21 days of test

PPCP (µg L⁻¹)	*C. haspan*				*H. coronarium*				*H. psittacorum*			
	Initial	Day 7	Day 14	Day 21	Initial	Day 7	Day 14	Day 21	Initial	Day 7	Day 14	Day 21
0	0	0	0	0	0	0	0	0	0	0	0	0
10	0	0	0	0	0	0	0	0	0	0	0	0
100	0	1	1	1	0	1	1	2	0	0	0	0
1000	0	1	1	1	0	1	3	3	0	1	2	3
5000	0	1	2	3	0	2	3	4	0	2	3	3
10000	0	1	3	3	0	1	4	4	0	3	4	4

The chlorosis values were determined as 'none' (0), 'low' (1), 'middle' (2), 'high' (3), and 'necrosis' (4).

Table 4 | Significant differences of biomass growth between plant species after 21 days of exposure to PPCPs

Plant	Groups	Significative differences between groups	Value significance – Mann Whitney
H. psittacorum (a)	a – b		0.606
	a – c		0.110
H. coronarium (b)	b – a		0.606
	b – c	*	0.006
C. haspan (c)	c – a		0.110
	c – b	*	0.006

*Significative differences (*P*<0.05).

Table 5 | Significant differences of PPCPs concentrations over biomass growth after 21 days exposure

Concentration (μg L^{-1})	Groups	Significant differences between groups	Value significance – Mann Whitney
10,000 (a)	a – b		0.931
	a – c	*	0.019
	a – d	*	0.000
	a – e	*	0.000
	a – f	*	0.000
5,000 (b)	b – a		0.931
	b – c	*	0.014
	b – d	*	0.000
	b – e	*	0.000
	b – f	*	0.000
1,000 (c)	c – a	*	0.014
	c – b	*	0.019
	c – d		0.113
	c – e		0.113
	c – f	*	0.050
100 (d)	d – a	*	0.000
	d – b	*	0.000
	d – c		0.113
	d – e		0.436
	d – f	*	0.024
10 (e)	e – a	*	0.000
	e – b	*	0.000
	e – c		0.113
	e – d		0.436
	e – f		0.430
0 (f)	f – a	*	0.000
	f – b	*	0.000
	f – c	*	0.050
	f – d	*	0.024
	f – e		0.340

*Significative differences (*P*<0.05).

3.2. Stage II – removal

Figure 4 shows the removal curves of the compounds acetylsalicylic acid, ketoprofen, and tonalide from the nutrient solution. The removal of acetylsalicylic acid in the first 24 h reaches 75% for *H. psittacorum* and 84% for *C. haspan*, biodegradation, and photodegradation reactors; given that all reactors with final concentrations below 3.9 μg L^{-1} (quantification limit of the analytic method employed), it is hard to determine the main removal mechanism, however, based on the evidence reported in previous studies, its removal it may be specially attributed to the degradation by biological processes (Hijosa-Valsero *et al.*

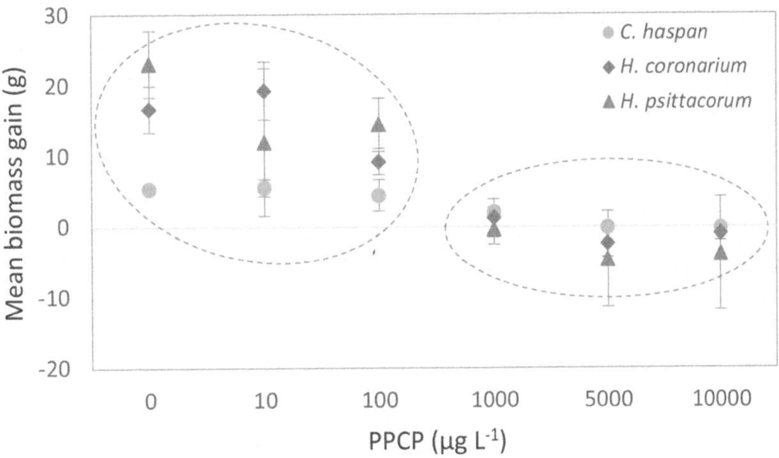

Figure 3 | Mean biomass gain of tested plant species as function of PPCP exposure. Dotted circles indicate main concentrations groups based on a hierarchical conglomerate analysis. Bars indicated standard deviation of the mean.

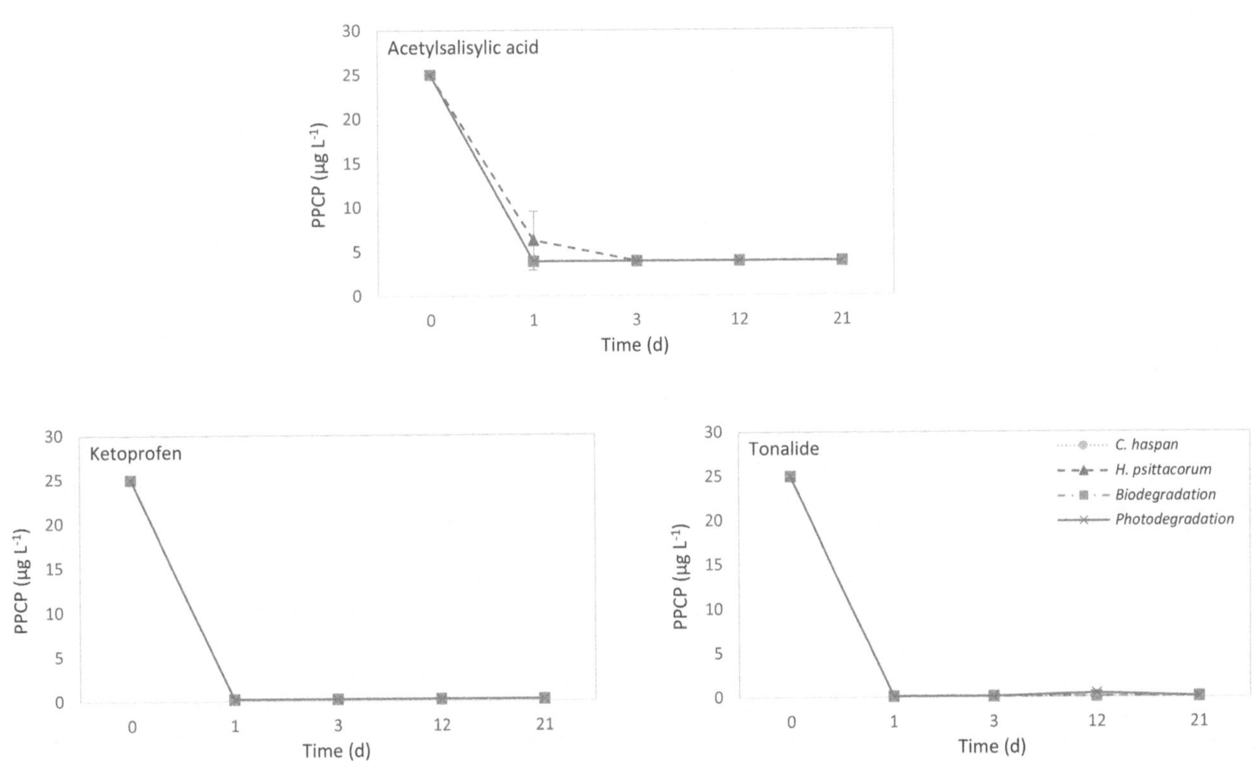

Figure 4 | Removal curves of acetylsalicylic acid, ketoprofen, and tonalide by phytoremedation (*C. haspan* and *H. psittacorum*), biodegradation, and photodegradation mechanisms (Stage II). Bars indicate standard deviation of the mean.

2010; Zhang *et al.* 2012a; Li *et al.* 2014). The compounds ketoprofen and tonalide were 100% removed during the first 24 hours, and as in the case of acetylsalicylic acid, it is not possible to distinguish the main elimination pathway given the quick removal of the mentioned compounds. However, several studies consider that the photolytic degradation is the main mechanism for the decrease in the recalcitrant compounds such as ketoprofen (Hijosa-Valsero *et al.* 2010; Matamoros & Salvadó 2012; Reyes *et al.* 2012; Zhang *et al.* 2012a; Zhang *et al.* 2014), whereas tonalide is a lipophilic compound that tends to adhere to solids on different surfaces (Ávila *et al.* 2010; Hijosa-Valsero *et al.* 2010), with sorption the main elimination path due to its hydrophobic character and high K_{ow} of 3.933 (Zhang *et al.* 2014).

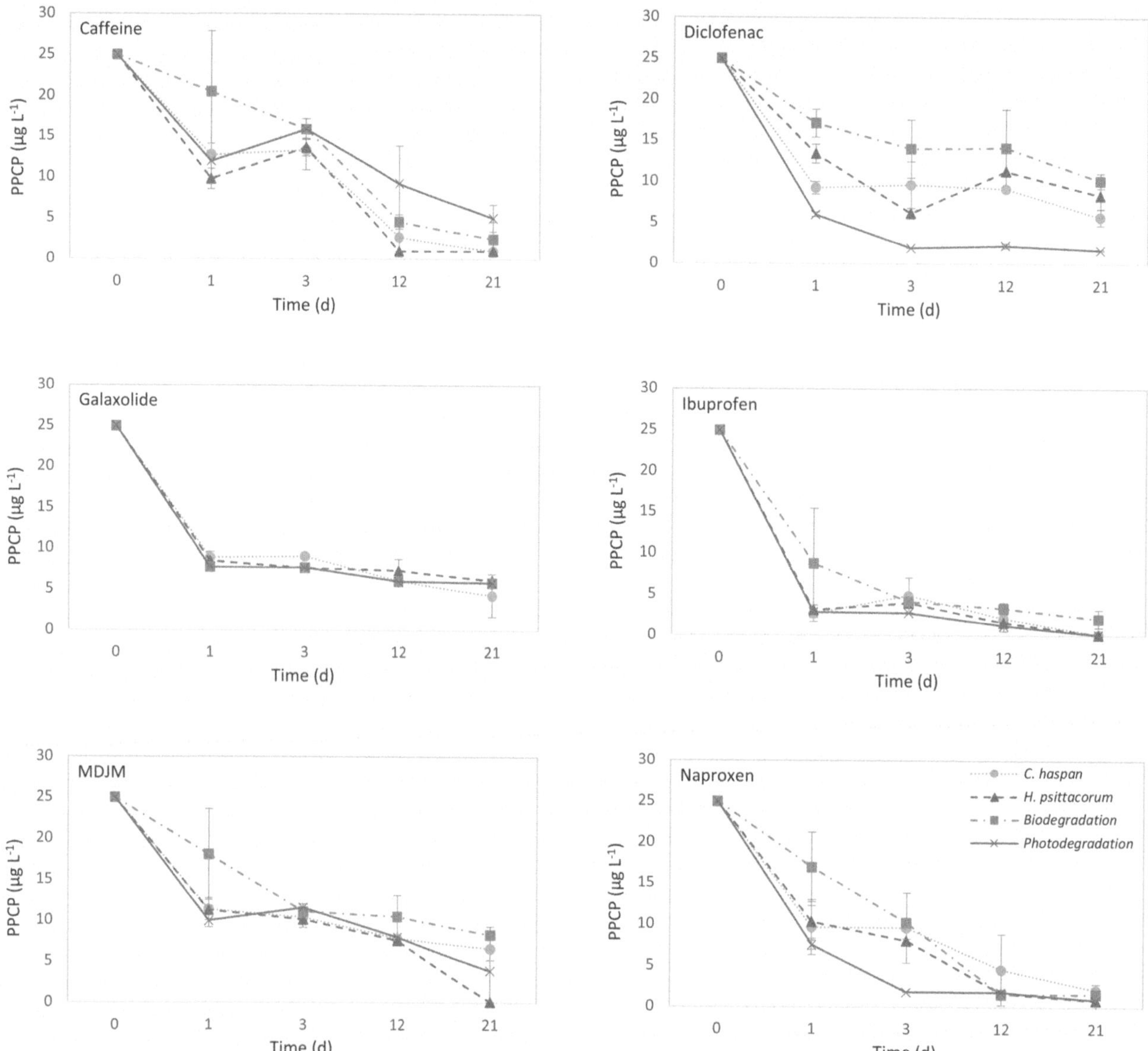

Figure 5 | Removal curves of caffeine, diclofenac, galaxolide, ibuprofen, MDJM, and naproxen by phytodegradation (*C. haspan* and *H. psittacorum*), biodegradation, and photodegradation mechanisms (Stage II). Bars indicate standard deviation.

Removal curves of the compounds caffeine, diclofenac, galaxolide, ibuprofen, naproxen, and methyl dihydrojasmonate (MDJM) are presented in Figure 5. Caffeine is removed 96% in the planted reactors, 90% in the covered reactors, and 80% in the reactors exposed to light, suggesting that the main removal mechanism is given by the action of plants; the obtained results are comparable with those reported by Zhang *et al.* (2013a) and Matamoros *et al.* (2012), who in planted hydroponic reactors reported caffeine removals of 99%, being the absorption by plants the main removal mechanism, and in a lesser proportion by photocatalytic action and microbial degradation. The elimination of caffeine by the action of plants is mainly associated to its ionizable character, polarity and high solubility in water, characteristics that allow it to be absorbed by the plant and translocated to the plant's organs (Li *et al.* 2014).

Diclofenac is removed in 93% by photolytic action and in a lesser proportion by phytodegradation (*H. psittacorum* – 66%; *C. haspan* – 79%), and biodegradation (59%); Zhang *et al.* (2012b) found that Diclofenac may be eliminated in 80% by photodegradation, however the plant absorption also represents a passive elimination mechanism of this compound due to its high hydrophobicity; Matamoros *et al.* (2012) also reported removals superior to 99% on reactors exposed to light, classifying Diclofenac as a compound highly photodegradable.

Galaxolide is removed 76% in *H. psittacorum* reactors, 83% in *C. haspan* reactors, 77% by photodegradation, and 65% by biodegradation; these results are similar to those found in other studies (Hijosa-Valsero *et al.* 2010; Carranza *et al.* 2014) that report removals between 0 and 80%, however it is hard to establish which is the predominant removal mechanism in the assessed reactors. The removal efficiency of ibuprofen at the end of the experiment was 99% for planted and photodegradation reactors, whereas biodegradation reactors reached removal efficiencies of 92%. The degradation curve of this compound indicates that the main removal mechanisms were phytodegradation and photodegradation; similar results were found by Hijosa-Valsero *et al.* (2010), who established that the removal of ibuprofen is given under a metabolic aerobic mechanism, and the elimination is favored by the presence of plants, attributing this fact to the modification on the redox values in the roots. Zhang *et al.* (2011) also reported high removal efficiencies of ibuprofen on planted systems, attributing its elimination to low K_{ow} and the rhizosphere action that favors an oxidant environment, favoring microbial activity; Dordio *et al.* (2010) stated that biodegradation is the main elimination mechanism due to lipophile moderated by its low K_{wo} that benefits movement through cellular membranes and inclusion into the transpiration plant's root.

Naproxen is removed more than 90% in all the reactors for the entire period of the test, and the relationship between exposition and elimination times, indicate that photodegradation is the main elimination mechanism; these results are coherent with the ones reported by Zhang *et al.* (2013b), who obtained removals of more than 90% in the reactors exposed to the sunlight, thus indicating the importance of the photocatalytic action in the elimination of naproxen. At the same time, Cardinal *et al.* (2014) indicated that direct photolysis represents 100% elimination of naproxen.

MDJM is 99% removed in reactors planted with con *H. psittacorum*, 74% in reactors planted with *C. haspan*, 84% in reactors exposed to light, and 66% in the covered reactors without plants. These results are coherent with previous studies where MDJM is exposed as a hydrophilic substance easily degradable with removal efficiencies of relatively high masses (>80%) mainly as a result of its high biodegradation rate (Matamoros & Salvadó 2012), and where the presence of plants highly contributes in its elimination (Hijosa-Valsero *et al.* 2010).

The ANOVA and Tukey HSD analysis of removal rates at the end of the experiment (Table 6), confirm the results of Stage II, indicating that for certain compounds (MDJM, diclofenac, and ibuprofen) there are significant differences by type of reactor ($p < 0.05$).

3.3. Incidence of macrophytes and photodegradation mechanisms in the removal of PPCPs

In Figure 6 is summarized the removal rate of the PPCP compounds, which are classified in two big blocks in agreement with the tested elimination mechanism in each reactor: block 1, where easily absorbed compound are found and eliminated by the action of plants, and a second block formed by compounds eliminated by photocatalytic action. Plants play a significant role in the direct absorption of many contaminants, even when not having the support of specific drivers for absorption of xenobiotic organic compounds, making the PPCPs inside plants being driven by diffusion according with the hydrophobicity of each compound (Li *et al.* 2014).

The results of the experiment indicated that the compounds ibuprofen, caffeine, MDJM and galaxolide are removed in a more efficient way in planted reactors. The compound MDJM is principally removed by *H. psittacorum*, while the galaxolide is more efficiently removed by *C. haspan*. Caffeine and ibuprofen presented the same elimination rate in both species. Even though there is not a clear differentiation in the removal mechanism for acetylsalicylic acid and tonalide, several authors

Table 6 | Tukey HSD least square means of removal rates (%) of tested PPCPs after 21 days of exposure

Reactor	Acetylsalicylic acid	Caffeine	MDJM	Diclofenac	Galaxolide	Ibuprofen	Ketoprofen	Naproxen	Tonalide
C. haspan	84.40[a]	96.00[a]	73.60[b]	77.20[ab]	83.00[a]	99.80[a]	99.05[a]	91.60[a]	99.60[a]
H. psittacorum	84.40[a]	96.00[a]	99.60[a]	66.60[bc]	75.80[a]	99.60[a]	99.05[a]	96.80[a]	99.60[a]
Biodegradation	84.3[a]	90.00[a]	67.00[b]	59.40[c]	65.20[a]	91.80[a]	99.20[a]	93.80[a]	99.45[a]
Photodegradation	84.3[a]	79.80[a]	69.60[b]	93.40[a]	76.80[a]	99.60[a]	99.20[a]	96.60[a]	99.45[a]
Pr>F(Model)	0.058	0.104	0.001	0.005	0.236	0.047	0.058	0.423	0.058
Significant	No	No	Yes	Yes	No	Yes	No	No	No

Superscript letters indicate significant differences among reactors.

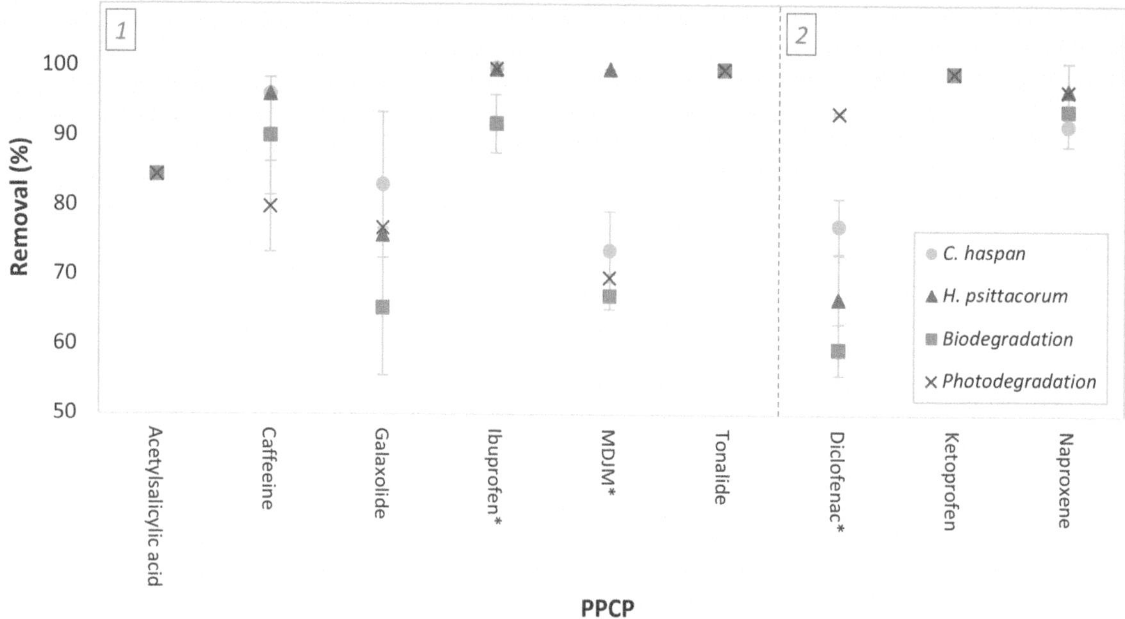

Figure 6 | Removal rate of PPCPs after 21 days of exposure, classified by removal mechanisms. Block 1: compounds prone to phytodegradation; Block 2: compounds prone to photodegradation. Bars indicate standard deviation. * Indicates PPCPs with significant removal rates on tested systems ($P<0.05$; ANOVA).

conclude that their removal is mainly due to the action of the plants (Ávila *et al.* 2010; Hijosa-Valsero *et al.* 2010; Zhang *et al.* 2012a; Li *et al.* 2014).

The photolysis occurs when the compounds absorb photons to form oxygen reactive species (Cardinal *et al.* 2014). In this experiment, around 90% of diclofenac is removed mainly in reactors exposed to the sunlight. Similarly, naproxen is removed mainly by the photolytic action and in lesser way by the action of the *H. psittacorum*. On the other hand, the elimination rates of ketoprofen did not allow fairly distinguishing between the main elimination paths, but the literature reviewed highlights photodegradation as the main removal mechanism (Hijosa-Valsero *et al.* 2010; Matamoros & Salvadó 2012; Reyes *et al.* 2012; Zhang *et al.* 2012a; Zhang *et al.* 2014).

4. CONCLUSIONS

The toxicity response to PPCPs in the tested plants is reflected in the reactors with concentrations up to 1,000 µg L^{-1}; above that concentration evidence was found of growth inhibition, loss of biomass, and necrosis, forming two main groups according to concentration and impact on plants. A first group corresponding to the absence of toxic responses for concentration inferior to 1,000 µg L^{-1}, and a second group that reports toxicity from 5,000 µg L^{-1}. In this form the species *H. psittacorum and C. haspan* are distinguished as the most tolerant and suitable for the removal of PPCPs.

The highest removal of the assessed compounds occurred during the first 24 hours of experimentation, thus identifying two blocks of PPCP elimination: one block formed by the compounds acetylsalicylic acid, caffeine, galaxolide, ibuprofen, MDJM, and tonalide that could be removed in more than 90% by the action of the plants or their exudates, and second block formed by the substances diclofenac, ketoprofen, and naproxen that are mainly eliminated by photolytic action. The results allow the conclusion that the presence of macrophytes in the assessed reactors favor the removal of studied PPCPs and show an interesting potential to be implemented in real-scale wetland systems to treat that kind of pollutant in tropical areas.

Further studies should be carried out using lower concentrations of the tested compounds, such as the ones found in wastewaters, as well as in natural waters. Also, to complement the results of this study, it is recommended to carry out further studies to identify intermediate metabolites derived of the degradation of the studied compounds, as well as the analysis of to corroborate the bioaccumulation and/or transformation potential of tested PPCPs in different plant tissues.

ACKNOWLEDGEMENTS

Funding for this work was provided by Colciencias (Science, technology, and innovation department in Colombia), the World Bank, and the Universidad Tecnológica de Pereira. We also thank staff of Water and Sanitation Research Group for assistance with the experimental setup and sampling.

DATA AVAILABILITY STATEMENT

All relevant data are included in the paper or its Supplementary Information.

REFERENCES

Addinsoft 2021 *XLSTAT Statistical and Data Analysis Solution*.

Alkhatib, R., Bsoul, E., Blom, D. A., Ghoshroy, K., Creamer, R. & Ghoshroy, S. 2013 El análisis microscópico de la acumulación de plomo en el tabaco *(Nicotiana tabacum var. Turcas)* raíces y hojas (The microscope analysis of lead accumulation in tobacco *(Nicotiana tabacum var. Turcas)* roots and leaves). Diario de Microscopía y Ultraestructura, Vol. 1 (1-2), 57–62.

APHA 2012 *Standard methods for the examination of water and wastewater*, 22nd edn. Washington, DC, USA, p. 1360, ISBN: 978-087553-013-0.

Ávila, C., Pedescoll, A., Matamoros, V., Bayona, J. M. & García, J. 2010 Capacity of a horizontal subsurface flow constructed wetland system for the removal of emerging pollutants: an injection experiment. *Chemosphere* **81** (9), 1137–1142. https://doi.org/10.1016/j.chemosphere.2010.08.006.

Ávila, C., Reyes, C., Bayona, J. M. & García, J. 2013 Emerging organic contaminant removal depending on primary treatment and operational strategy in horizontal subsurface flow constructed wetlands: influence of redox. *Water Research* **47** (1), 315–325. https://doi.org/10.1016/j.watres.2012.10.005.

Bolong, N., Ismail, A. F., Salim, M. R. & Matsuura, T. 2009 A review of the effects of emerging contaminants in wastewater and options for their removal. *Desalination* **239** (1–3), 229–246. https://doi.org/10.1016/j.desal.2008.03.020.

Brunhoferova, H., Venditti, S., Schlienz, M. & Hansen, J. 2021 Removal of 27 micropollutants by selected wetland macrophytes in hydroponic conditions. *Chemosphere* **281** (December 2020), 130980. https://doi.org/10.1016/j.chemosphere.2021.130980.

Cardinal, P., Anderson, J. C., Carlson, J. C., Low, J. E., Challis, J. K., Beattie, S. A., Bartel, C. N., Elliott, A. D., Montero, O. F., Lokesh, S., Favreau, A., Kozlova, T. A., Knapp, C. W., Hanson, M. L. & Wong, C. S. 2014 Macrophytes may not contribute significantly to removal of nutrients, pharmaceuticals, and antibiotic resistance in model surface constructed wetlands. *Science of the Total Environment* **482–483**, 294–304. https://doi.org/10.1016/j.scitotenv.2014.02.095.

Carranza-Diaz, O., Schultze-Nobre, L., Moeder, M., Nivala, J., Kuschk, P. & Koeser, H. 2014 Removal of selected organic micropollutants in planted and unplanted pilot-scale horizontal flow constructed wetlands under conditions of high organic load. *Ecological Engineering* **71**, 234–245. https://doi.org/10.1016/j.ecoleng.2014.07.048.

Deblonde, T., Cossu-Leguille, C. & Hartemann, P. 2011 Emerging pollutants in wastewater: a review of the literature. *International Journal of Hygiene and Environmental Health* **214** (6), 442–448. https://doi.org/10.1016/j.ijheh.2011.08.002.

Dordio, A. V. & Carvalho, A. J. P. 2013 Organic xenobiotics removal in constructed wetlands, with emphasis on the importance of the support matrix. *Journal of Hazardous Materials* **252–253**, 272–292. https://doi.org/10.1016/j.jhazmat.2013.03.008.

Dordio, A., Palace Carvalho, A. J., Martins Teixeira, D., Barrocas Dias, C. & Pinto, A. P. 2010 Removal of pharmaceuticals in microcosm constructed wetlands using Typha spp. and LECA. *Bioresource Technology* **101** (3), 886–892. https://doi.org/10.1016/j.biortech.2009.09.001.

Hijosa-Valsero, M., Matamoros, V., Sidrach-Cardona, R., Martín-Villacorta, J., Bécares, E. & Bayona, J. M. 2010 Comprehensive assessment of the design configuration of constructed wetlands for the removal of pharmaceuticals and personal care products from urban wastewaters. *Water Research* **44** (12), 3669–3678. https://doi.org/10.1016/j.watres.2010.04.022.

Iori, V., Pietrini, F. & Zacchini, M. 2012 Assessment of ibuprofen tolerance and removal capability in *Populus nigra* L. by in vitro culture. *Journal of Hazardous Materials* **229–230**, 217–223. https://doi.org/10.1016/j.jhazmat.2012.05.097.

Li, Y., Zhu, G., Jern Ng, W. & Tan, S. K. 2014 A review on removing pharmaceutical contaminants from wastewater by constructed wetlands: design, performance and mechanism. *Science of the Total Environment* **468–469**, 908–932. https://doi.org/10.1016/j.scitotenv.2013.09.018.

Liu, J.-L. & Wong, M.-H. 2013 Pharmaceuticals and personal care products (PPCPs): a review on environmental contamination in China. *Environment International* **59**, 208–224. https://doi.org/10.1016/j.envint.2013.06.012.

Matamoros, V. & Bayona, J. M. 2006 Elimination of pharmaceuticals and personal care products in subsurface flow constructed wetlands. *Environmental Science and Technology* **40** (18), 5811–5816. https://doi.org/10.1021/es0607741.

Matamoros, V. & Salvadó, V. 2012 Evaluation of the seasonal performance of a water reclamation pond-constructed wetland system for removing emerging contaminants. *Chemosphere* **86** (2), 111–117.

Matamoros, V., Arias, C. & Brix, H. 2007 Removal of pharmaceuticals and personal care products (PPCPs) from urban wastewater in a pilot vertical flow constructed wetland and a sand filter. *Environmental Science and Technology* **41** (23), 8171–8177. https://doi.org/10.1021/es071594+.

Matamoros, V., Caselles-Osorio, A., García, J. & Bayona, J. M. 2008 Behaviour of pharmaceutical products and biodegradation intermediates in horizontal subsurface flow constructed wetland. A microcosm experiment. *Science of the Total Environment* **394** (1), 171–176. https://doi.org/10.1016/j.scitotenv.2008.01.029.

Matamoros, V., Arias, C., Brix, H. & Bayona, J. M. 2009 Preliminary screening of small-scale domestic wastewater treatment systems for removal of pharmaceutical and personal care products. *Water Research* **43** (1), 55–62. https://doi.org/10.1016/j.watres.2008.10.005.

Matamoros, V., Xuan, N. L., Arias, C. A., Salvadó, V. & Brix, H. 2012 Evaluation of aquatic plants for removing polar microcontaminants: a microcosm experiment. *Chemosphere* **88** (10), 1257–1264. https://doi.org/10.1016/j.chemosphere.2012.04.004.

Mompelat, S., Le, B. B. & Thomas, O. 2009 Occurrence and fate of pharmaceutical products and by-products, from resource to drinking water. *Environment International* **35** (5), 803–814. https://doi.org/10.1016/j.envint.2008.10.008.

Nadeem, S. M., Ahmad, M., Ahmad, Z. Z., Javaid, A. & Ashraf, M. 2014 The role of mycorrhizae and plant growth promoting rhizobacteria (PGPR) in improving crop productivity under stressful environments. *Biotechnology Advances* **32** (2), 429–448. https://doi.org/10.1016/j.biotechadv.2013.12.005.

Ortiz de García, S., Pinto, G., García Encina, P. & Irusta Mata, R. 2013 Consumption and occurrence of pharmaceutical and personal care products in the aquatic environment in Spain. *Science of the Total Environment* **444**, 451–465. https://doi.org/10.1016/j.scitotenv.2012.11.057.

Reyes-Contreras, C., Hijosa-Valsero, M., Sidrach-Cardona, R., Bayona, J. M. & Bécares, E. 2012 Temporal evolution in PPCP removal from urban wastewater by constructed wetlands of different configuration: a medium-term study. *Chemosphere* **88** (2), 161–167. https://doi.org/10.1016/j.chemosphere.2012.02.064.

Santos Lúcia, H. M. L. M., Araújo, A. N., Fachini, A., Pena, A., Delerue-Matos, C. & Montenegro, M. C. B. S. M. 2010 Ecotoxicological aspects related to the presence of pharmaceuticals in the aquatic environment. *Journal of Hazardous Materials* **175** (1–3), 45–95. https://doi.org/10.1016/j.jhazmat.2009.10.100.

Sui, Q., Huang, J., Deng, S., Yu, G. & Fan, Q. 2010 Occurrence and removal of pharmaceuticals, caffeine and DEET in wastewater treatment plants of Beijing, China. *Water Research* **44** (2), 417–426. https://doi.org/10.1016/j.watres.2009.07.010.

US Environmental Protection Agency (USEPA) 2002 *Frequently Asked Questions: PPCPs as Environmental Pollutants*. Available from: http://www.epa.gov/esd/chemistry/pharma/faq.html.

Verlicchi, P., Galletti, A., Petrovic, M., Barceló, D., Al Aukidy, M. & Zambello, E. 2013 Removal of selected pharmaceuticals from domestic wastewater in an activated sludge system followed by a horizontal subsurface flow bed – analysis of their respective contributions. *Science of the Total Environment* **454–455**, 411–425. https://doi.org/10.1016/j.scitotenv.2013.03.044.

WHO – World Health Organization 2012 *Pharmaceuticals in Drinking Water*. Geneva, Switzerland. ISBN: 978 9241502085.

Yang, X., Flowers, R. C., Weinberg, H. S. & Singer, P. C. 2011 Occurrence and removal of pharmaceuticals and personal care products (PPCPs) in an advanced wastewater reclamation plant. *Water Research* **45** (16), 5218–5228. https://doi.org/10.1016/j.watres.2011.07.026.

Zhang, D. Q., Tan, S. K., Gersberg, R. M., Sadreddini, S., Zhu, J. & Tuan, N. A. 2011 Removal of pharmaceutical compounds in tropical constructed wetlands. *Ecological Engineering* **37** (3), 460–464. https://doi.org/10.1016/j.ecoleng.2010.11.002.

Zhang, D. Q., Gersberg, R. M., Zhu, J., Hua, T., Jinadasa, K. B. S. N. & Tan, S. K. 2012a Batch versus continuous feeding strategies for pharmaceutical removal by subsurface flow constructed wetland. *Environmental Pollution* **167**, 124–131. https://doi.org/10.1016/j.envpol.2012.04.004.

Zhang, D. Q., Hua, T., Gersberg, R. M., Zhu, J., Jern, N. W. & Tan, S. K. 2012b Fate of diclofenac in wetland mesocosms planted with *Scirpus validus*. *Ecological Engineering* **49**, 59–64. https://doi.org/10.1016/j.ecoleng.2012.08.018.

Zhang, D. Q., Hua, T., Gersberg, R. M., Zhu, J., Jern, N. W. & Tan, S. K. 2013a Fate of caffeine in mesocosms wetland planted with *Scirpus validus*. *Chemosphere* **90** (4), 1568–1572. https://doi.org/10.1016/j.chemosphere.2012.09.059.

Zhang, D. Q., Hua, T., Gersberg, R. M., Zhu, J., Jern, N. W. & Tan, S. K. 2013b Carbamazepine and naproxen: fate in wetland mesocosms planted with *Scirpus validus*. *Chemosphere* **91** (1), 14–21. https://doi.org/10.1016/j.chemosphere.2012.11.018.

Zhang, D. Q., Gersberg, R. M., Jern, N. W. & Keat, T. S. 2014 Removal of pharmaceuticals and personal care products in aquatic plant-based systems: a review. *Environmental Pollution* **184**, 620–639. https://doi.org/10.1016/j.envpol.2013.09.009.

Zhang, Y., Lv, T., Carvalho, P. N., Arias, C. A., Chen, Z. & Brix, H. 2015 Removal of the pharmaceuticals ibuprofen and iohexol by four wetland plant species in hydroponic culture: plant uptake and microbial degradation. *Environmental Science and Pollution Research*. January 2016. https://doi.org/10.1007/s11356-015-5552-x

First received 16 December 2021; accepted in revised form 11 March 2022. Available online 22 March 2022

doi: 10.2166/wst.2022.075

Influence of plants on the salinity of constructed wetlands effluents

Denis Leocádio Teixeira

Institute of Agricultural Sciences, Federal University of Jequitinhonha and Mucuri Valleys, Avenida Universitária, no 1.000, Bairro Universitários, Unaí 38610-000, Brazil
E-mail: denis.teixeira@ufvjm.edu.br

DLT, 0000-0001-7343-9696

ABSTRACT

Reuse of constructed wetlands effluents for fertigation is a promising practice in regions that suffer from water scarcity. Thus this study aimed to evaluate the influence of Vetiver (*Chrysopogon zizanioides*) and Tifton 85 (*Cynodon* sp.) grasses cultivation on salinity of the saturated solution of the Horizontal Subsurface Flow Constructed Wetlands (HSSF-CW) porous medium. The grasses were cultivated in HSSF-CW mesocosm and submitted to different levels of salinity for six months. The crop's evapotranspiration, biomass productivity, the electrical conductivity variation (ΔEC) in the solution under treatment and the water use efficiency (WUE) of these grasses were evaluated. The cultivation of Vetiver and Tifton 85 in HSSF-CW provided an increase in the electrical conductivity of the solution mainly for the lowest levels of affluent salinity. The ΔEC in HSSF-CW grown with Vetiver and Tifton 85 grasses were 90.20 and 27.13 mS cm^{-1}, respectively. Meteorological variables maximize the effect of the vegetative development in ΔEC of the solution. The WUE values of Vetiver and Tifton 85 grasses were 1.77 and 4.18 g kg^{-1}, respectively. Thus plants with high water use efficiency such as Tifton 85 is indicated for HSSF-CW in which the effluents are used in the fertigation of agricultural crops.

Key words: evapotranspiration, salinity, wastewater, water use efficiency

HIGHLIGHTS

- Cultivation of grass in HSSF-CW provided an increase in effluent EC.
- WUE is an essential metric in selecting species for cultivation in HSSF-CW.
- Evapotranspiration and productivity have a great influence on effluent reuse.
- HSSF-CW fed with weakly saline solutions provided greater relative ΔEC.

INTRODUCTION

Horizontal Subsurface Flow Constructed Wetlands (HSSF-CW) have been used to treat several types of wastewater, due to their simplicity, low cost, and ease of operation and maintenance. These treatment systems have as the main objective to reduce the organic load and ion content present in the wastewater (Kadlec & Wallace 2009; Fia *et al.* 2014; Tuttolomondo *et al.* 2015; Almeida *et al.* 2017; Liang *et al.* 2017; Saraiva *et al.* 2019).

In semiarid regions that suffer from water scarcity, water reuse is a real necessity, consequently, it is increasingly common sewage treatment plants use effluents for different purposes (Lavrnié *et al.* 2017). Constructed wetlands effluents are often used for fertigation of crops, supplying water and nutrients of the plants (Nan *et al.* 2020). However, to be a sustainable practice, the effluent must show low salinity and not compromise the receiving soil's physical structure and not cause phytotoxicity to cultivated plants (Ayers & Westcot 1985). High evapotranspiration rates can contribute to the effluent salinity increase of the HSSF-CW, making unfeasible to reuse, especially in regions with a hot and dry climate (Headley *et al.* 2012; Nan *et al.* 2020).

Wastewaters present a high variety of ions in their composition, which can be solubilized or constituted in the organic matter in suspension. Thus, when flowing through the HSSF-CW, the effluent salinity can be higher or lower than the affluent depending on several factors. The decrease in the ion concentration occurs due to absorption by the plants, microorganism's immobilization, insoluble salts precipitation, rain, and even volatilization. On the other hand, the increase has as precedent the organic matter mineralization, that makes ions available in the solution, or the evapotranspiration, which is responsible for the ion concentration in the effluent (Fia *et al.* 2014; Gao *et al.* 2014; Liang *et al.* 2017).

According to Kadlec & Wallace (2009), water lost to the atmosphere by plants' evaporation and transpiration, a process known as evapotranspiration, is the main factor influencing the salinity variation in the constructed wetlands systems. According to the authors, the salinity concentration, measured by the solution's electrical conductivity, precisely indicates the effect of dilution or concentration due to, respectively, precipitation and evapotranspiration observed in the HSSF-CW. According to Tuttolomondo *et al.* (2016) the evapotranspiration, in hot climate places, presents higher influence in the dynamics solute when compared to precipitation.

Several studies have been done with the objective to evaluate the cultivation of halophytic plants in constructed wetlands for the reduction in the salt concentration in wastewater. Freedman *et al.* (2014), when evaluating the salt absorption by plants and the crop evapotranspiration, concluded that these two processes occur simultaneously, in such way that the increase or decrease of the effluent salinity depends on the relative magnitude of each of these two processes. The authors observed that the increase in salinity in an HSSF-CW cultivated with *B. indica*, was due to high evapotranspiration in this system. In another study done by Fountoulakis *et al.* (2017), the phytoremediation potential of halophytic plants was enough to circumvent the increase in salinity due to evapotranspiration, resulting in an unchanged effluent electrical conductivity from the HSSF-CW. In the other hand, Shelef *et al.* (2012) verified a reduction in the wastewater salinity treated in vertical constructed wetlands, cultivated with *B. Indica*. According to the authors, the ion absorption by the plants, especially sodium and potassium, was responsible for the electrical conductivity reduction in the effluents from the treatment system.

Most of the available studies in the literature correlate salinity, plant development, treatment efficiency, and crop evapotranspiration. However, there is no precise information of the intervening factors in the wastewater electrical conductivity variation treated from HSSF-CW. The research's results show the necessity of more investigations to optimize the phytoremediation effect of plants in this treatment system (Freedman *et al.* 2014; Fountoulakis *et al.* 2017). According to Liang *et al.* (2017), wastewaters present different compositions and ionic concentrations, organic matters, beyond other pollutant elements that can interact and influence the ion absorption by plants and its adsorption to substrate of the HSSF-CW, making the evaluation of factors that interfere in the effluents' salinity difficult.

Considering that, this work had as the objective to evaluate the influence of the cultivation of the grasses Vetiver (*Chrysopogon zizanioides*) and Tifton 85 (*Cynodon* sp.) in electrical conductivity of the saturating solution of the HSSF-CW porous medium. Specifically, to evaluate the influence of evapotranspiration, the plant biomass productivity and climatic conditions, in electrical conductivity variation of the solution under treatment.

MATERIALS AND METHODS

This experiment was conducted at the Hydraulics, Irrigation, and Drainage Experimental Area from the Agricultural Engineering Department at the Federal University of Viçosa, Minas Gerais, Brazil. The geographic coordinates are latitude 20°46′08″ S and longitude 42°51′44″ W, with average sea level of 674 m. The region climate is defined as Cwa (Koppën), mesothermic with rainy summers and dry winters. The average annual rainfall is 1,221 mm and the average annual temperature varies between 19.0 and 20.0 °C.

Grass cultivation in HSSF-CW mesocosm

The influence evaluation from Vetiver and Tifton 85 grasses in the electrical conductivity variation (ΔEC) of the porous medium saturating solution was realized from the cultivation of these plants in HSSF-CW mesocosm for a period of six months (May 18, 2016–November 18, 2016). The experiment was installed in a randomized block design (RBD), with six repetitions, with the objective to evaluate the effect of the two grasses and the six affluent salinity levels over the ΔEC.

The HSSF-CW were built in masonry, with a rectangular shape, and dimensions of 0.92 m in length, 0.73 m in width, and 0.35 m in height. The system half support had a 0.3 m layer, filling with gneiss gravel #0, with D_{60} of 7.0 mm, uniformity coefficient of 1.6 (D_{60}/D_{10}) and 48.2% porosity. The saturation height was keeping of 0.25 m, and the average planting density of the Vetiver and Tifton 85 grasses was 12 propagules per m^2 (Figure 1).

The affluent salinity levels to each HSSF-CW mesocosm were monitored by the electrical conductivity (EC) of the nutrient solution, from the values of 0.2; 0.5; 1.0; 1.5: 2.0 and 3.0 dS m^{-1}. The utilization of the nutrient solution, in exchange of the wastewater, was intended to have higher control in the affluent EC during the treatment period because, as previously

Figure 1 | Construction of the HSSF-CW mesocosm (a) and the overview of the experimental site (b).

mentioned, the presence of organic matter and the high development of microorganisms in the HSSF-CW bed can hinder the observation of EC variations, due to factors related to the plant, climate, and wastewater variability.

The nutrient solution prepare was done separately for each pre-determined EC value, utilizing as reference the Hoagland and Arnon solution. The nutrients concentrations for the different system's affluent salinity levels are presented in Table 1.

The nutrient solution feeding for each mesocosm was made continuously, applying an average flow of 13.8 mL min^{-1} (20 L d^{-1}) through a dripper, resulting in a hydraulic retention time (HRT) of approximately three days. After each precipitation event, the total nutrient solution exchange was done by draining the porous medium solution of the mesocosm, once the objective of this work was to evaluate, mainly, the effect of evapotranspiration in the ΔEC of the treated solution, simulating hot and dry climate regions with low rainfall indices.

The effluent nutrient solution was totally collected and stored for the measurement of the hydric balance and the evapotranspiration determination of the culture (ET_C), as shown in Equation (1).

$$ET_C = \frac{(Q_a - Q_e)}{A} + P \tag{1}$$

where ET_C is culture evapotranspiration (mm d^{-1}); Q_a is affluent flow (L d^{-1}); Q_e is effluent flow (L d^{-1}); A is superficial area of the HSSF-CW mesocosm (m^2) and P is precipitation (mm d^{-1}).

The electrical conductivity variation (ΔEC) determination was performed by measuring the difference between the HSSF-CW mesocosm affluent and effluent EC using a bench conductivity meter, Hach brand, Sension 7 model. The obtained EC values in all measurements were converted to the temperature of 25 °C, as a way to standardize the results.

The cuts of the plant aerial parts were made every thirty days for determination of the biomass productivity, quantifying the dry weight, obtained after drying in air-forced oven at 65 °C for 48 h (Matos 2015). The air temperature, solar radiation, relative air humidity, wind speed, and precipitation data were collected daily in the automatic weather station installed at the experimental area.

Table 1 | Nutrients concentration for the different affluent salinity levels to the HSSF-CW mesocosm

EC (dS m^{-1})	N (mg L^{-1})	P (mg L^{-1})	K (mg L^{-1})	Ca (mg L^{-1})	Mg (mg L^{-1})	S (mg L^{-1})	Fe (mg L^{-1})	B (µg L^{-1})	Cu (µg L^{-1})	Mn (µg L^{-1})	Mo (µg L^{-1})	Zn (µg L^{-1})
0.2	11.5	0.8	6.5	10.4	3.1	4.1	0.3	32	1	45	1	6
0.5	38.3	3.8	24.3	25.9	7.8	10.4	0.8	81	3	113	2	15
1.0	82.8	8.9	53.9	51.8	15.6	20.7	1.6	162	6	227	3	29
1.5	127.4	13.9	83.6	77.8	23.3	31.1	2.4	243	10	340	5	44
2.0	172.0	19.0	113.2	103.7	31.1	41.5	3.2	324	13	454	6	58
3.0	261.1	29.0	172.4	155.5	46.7	62.2	4.9	486	19	680	10	87

Water use efficiency in HSSF-CW

The water use efficiency (WUE), defined as the amount of biomass produced by the unit of water consumed, is an indicator utilized to select species to be cultivated in HSSF-CW in arid regions. With the goal to share more information related to the cultivation of the Vetiver and Tifton 85 grasses in constructed wetlands, these indicators were determined during the experiments.

The WUEs of the two grasses were determined for each cut by dividing the dry plant biomass (g m^{-2}) by the total water volume lost by evapotranspiration (L m^{-2}), resulting in grams of aerial biomass per kilograms of consumed water (g kg^{-1}), considering the water specific weight equals to 998.2 g L^{-1} at 20 °C.

Statistical analysis of the data

The obtained data during the experimental period was statistically evaluated using analysis of variance (ANOVA). The differences of ΔEC values between plant species and affluent salinity levels were obtained by the Tukey test. Regression analyses were employed to obtain the functional relation between the affluent salinity levels to the HSSF-CW and the ΔEC, where the model's settings were evaluated by the F test, and the obtained parameters were submitted to the t-test. For all p-value interpretation, the threshold value was set to 0.05.

To evaluate the main intervening factors in electrical conductivity variation of the treated solution, Pearson correlation analyses were realized between the ΔEC and some climatic variables monitored during the experiment period. All statistical analyses were made using the *Statistica*® and *Microsoft Excel*® software.

RESULTS AND DISCUSSION

Climate conditions through the experimental period

The experiment comprised the autumn, winter, and spring seasons, thus, there was a reduction in temperature, precipitation, radiation, and evapotranspiration, from the beginning of the experiment until mid-June. After July, there was a gradual increase in the variables mentioned above, with some variations resulting from cloudy periods that caused a low solar radiation incidence on the surface. It is important to emphasize that precipitation did not compromise the results obtained during the experimental period since the dilution effect was eliminated by changing the nutrient solution.

In Figure 2, there is a high correlation between air temperature, solar radiation, and grass evapotranspiration since they are variables that act in parallel with water loss in HSSF-CW. Tifton 85 grass showed greater evapotranspiration than Vetiver grass, which may have been one of the factors that contributed to the differentiation of ΔEC in HSSF-CW, which will be discussed throughout this work. Warmer periods also contributed to the greater vegetative development of grasses, thus increasing water and nutrients' absorption by plants. The relative proportion between the absorption rates of water and nutrients can influence the ΔEC of the wastewater treated, therefore, plants that present high evapotranspiration and low

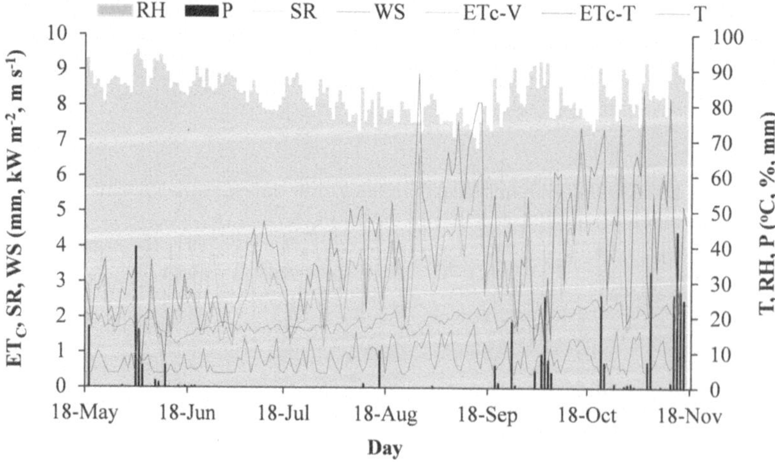

Figure 2 | Daily average values for relative humidity (RH), precipitation (P), solar radiation (SR), wind speed (WS), evapotranspiration of the Vetiver (ET$_C$-V) and Tifton (ET$_C$-T) grass, and air temperature (T) obtained during the experimental period.

productivity can contribute to the increase in effluent EC of the treatment system (Headley *et al.* 2012; Tuttolomondo *et al.* 2015; Nan *et al.* 2020).

Evaluation of salinity levels and grass species in the variation of EC

The effluent electrical conductivity of the HSSF-CW mesocosm showed different behavior between grass species and affluent salinity levels, however, on average, there was an increase in EC values after passing through the systems, resulting in positive ΔEC values.

ANOVA results indicated a significant effect of grass species ($p < 0.05$) and the levels of salinity affluent to the HSSF-CW ($p < 0.05$) in the ΔEC, however, the interaction between the factors was not significant ($p = 0.24$). Therefore, these factors act independently in the variation of the electrical conductivity of the saturating solution of the porous medium of the systems, being analyzed separately.

According to the results presented in Table 2, the cultivation of Vetiver grass in HSSF-CW provided higher values of effluent EC when compared to the cultivation of Tifton 85 grass. These results can be due to less removal of ions or greater water loss in the system grown with Vetiver grass. However, considering that Tifton 85 grass was the one that provided the greatest evapotranspiration (Figure 3), it is concluded that the Vetiver grass showed less ionic removal. Tifton 85, it seems, removed a higher number of ions in solution as a way to supply its nutritional demand. According to Saraiva *et al.* (2019), Tifton 85 has a high potential for sodium extraction in HSSF-CW due to its high biomass productivity.

Indeed, the Tifton 85 grass showed a higher plant biomass productivity than that obtained by Vetiver grass (Figure 3) ($p < 0.05$), resulting in greater nutritional demand for the vegetative development of this grass and, consequently, a higher rate of absorption of ions in solution.

The establishment of a simple and direct relationship between evapotranspiration and the increase in effluent EC of the HSSF-CW is not indicated when it is desired to compare systems cultivated with different plant species or submitted to different cultivation conditions. In these cases, one must consider the nutritional need of the plant, its vegetative development stage, climatic conditions, as well as the ionic composition of the wastewater treated. According to Freedman *et al.* (2014), plants submitted to different environmental and cultivation conditions can show completely different results in terms of increase or decrease in effluent EC of the treatment systems, depending on the relative magnitude of the evapotranspiration and ion absorption processes by the plants.

In HSSF-CW cultivated with Tifton 85, for example, higher affluent salinity levels provided a significant increase in the plant biomass productivity (Figure 3). However, crop evapotranspiration did not show significant variation, indicating that

Table 2 | Average variation of EC in the solution under treatment in HSSF-CW grown with Vetiver and Tifton 85 grasses

Treatment	ΔEC (mS cm^{-1})
Vetiver grass	90.20a
Tifton 85 grass	27.13b

Means followed by the same letter in the column, do not differ statistically from each other, at 5% significance, by the Tukey test.

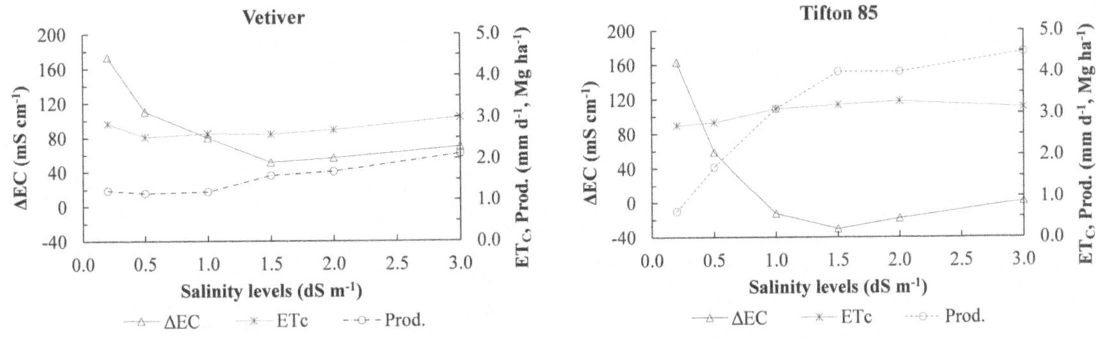

Figure 3 | Average values of electrical conductivity variation (ΔEC), evapotranspiration (ET$_C$) and plant biomass productivity (Prod.) for Vetiver and Tifton 85 grasses, as a function of affluent salinity levels.

plant ion uptake was the main responsible for electrical conductivity variation. The absorption of water by plants enhances the passive transport of nutrients (Taiz *et al.* 2017), thus, the higher the productivity, the higher the absorption of nutrients by the plants, causing a reduction in ΔEC.

Regarding the affluent salinity, it appears that the lowest levels were those that provided the greatest increase in effluent EC (Figure 3 and Table 3).

The solutions with less salinity have a lower concentration of essential nutrients to the plants (Table 1), resulting in less vegetative development and low ion absorption of the solution. The lower biomass productivity of the grasses cultivated in these HSSF-CW, associated with the greater evaporation of water, in proportion to the transpiration, was the main responsible for the highest elevation of the effluent EC. Due to the grass's low vegetative development, the HSSF-CW surface was more exposed to radiation, causing heating of the surface and increased evaporative flow. According to Bachand *et al.* (2014), there is a strong correlation between EC variation in flooded environments and evaporation/transpiration. The relative contribution of each evapotranspiration component can be estimated from models that use the water balance and the affluent and effluent EC to these systems.

Regarding the smallest variation in the solution EC observed for the highest levels of affluent salinity, Liang *et al.* (2017) found no significant difference between the values of affluent and effluent EC of constructed wetlands submitted to solutions containing a high concentration of nutrients. However, for solutions with low concentration, there was a high variation of the EC throughout the experimentation period. According to the authors, these results are related to the higher absorption of ions by plants grown in less concentrated solutions since the concentration effect resulting from evapotranspiration was suppressed with the replacement of deionized water to the treatment systems.

To obtain the functional relationship between the affluent salinity levels to the HSSF-CW and the ΔEC, regression analyses were performed. A model was obtained for each grass species, with the relative variation of EC in the treatment systems as a dependent variable (Figure 4).

According to the coefficient of determination (R^2) and the significance of the coefficients ($p < 0.05$), the decreasing exponential model adjusted well to the experimental data. Besides that, the referred model satisfactorily describes the behavior considered logical for the phenomenon under study since it is the widely used first-order kinetic equation. As previously discussed, HSSF-CW fed with weakly saline solutions provided greater relative ΔEC.

Table 3 | Average EC variation in solution treated in HSSF-CW subjected to different levels of affluent salinity

Treatment	ΔEC (mS cm^{-1})
0.2 dS m^{-1}	168.14a
0.5 dS m^{-1}	84.54b
3.0 dS m^{-1}	35.73bc
1.0 dS m^{-1}	33.66bc
2.0 dS m^{-1}	19.31c
1.5 dS m^{-1}	10.60c

Means followed by the same letter in the column, do not differ statistically from each other, at 5% significance, by the Tukey test.

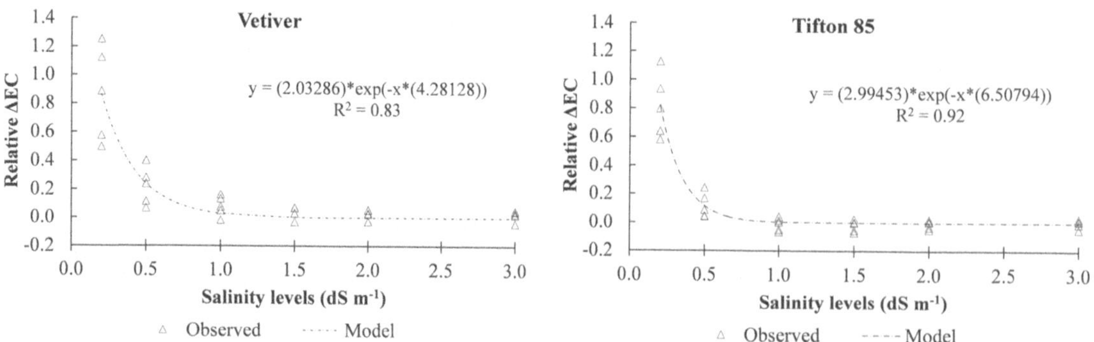

Figure 4 | Relative electrical conductivity variation curves (Relative ΔEC) as a function of the affluent salinity levels of the HSSF-CW.

For affluent salinity levels greater than 1.0 dS m^{-1}, it is observed that the relative ΔEC is approximately equal to zero; that is, the same HSSF-CW subjected to different climatic conditions presents a positive or negative variation of the EC depending on the month of cultivation. This demonstrates that the climatic factor is also associated with the EC effluent variation from the systems and not only the plant species and the wastewater salinity. According to Liang *et al.* (2017), ambient temperature and relative humidity greatly influence the treatment efficiency of the HSSF-CW when it comes to saline wastewater. According to the authors, these factors are decisive for the growth and development of plants grown in these systems, as they have significant relevance for photosynthesis and transpiration of crops.

The positive ΔEC values observed in the HSSF-CW grown with Vetiver grass (Figures 3 and 4), indicated that this plant's cultivation in the HSSF-CW could contribute to an increase in effluent EC, regardless of the ions' concentration present and climatic conditions. These results are corroborated by some authors who stated that Vetiver grass cultivation could result in an increase in the effluent EC of the HSSF-CW (Almeida *et al.* 2017; Silva *et al.* 2017). Regarding the Tifton 85 grass, there is no consensus regarding this effect: some authors observed an increase in effluent EC (Araujo *et al.* 2018; Saraiva *et al.* 2018), while others, a reduction (Fia *et al.* 2017; Jesus *et al.* 2020).

The results showed that the cultivation of plant species that present high evapotranspiration and low productivity of aerial plant biomass provide a greater potential for increasing the effluent EC of the HSSF-CW, which makes them crops not suitable for cultivation in arid environments where the reuse of the effluent is of utmost importance. For this assessment, the calculation of water use efficiency, which will be discussed in the next topic, becomes relevant.

Intervening factors in the variation of EC in HSSF-CW

To better understand the factors responsible for the variation in EC of the solution treated in HSSF-CW, correlation analysis were performed between ΔEC and ET$_C$ (mm d^{-1}), average air temperature (°C), relative humidity air (%), solar radiation (W m^{-2}) and wind speed (m s^{-1}), obtaining Pearson's correlation coefficient for the different levels of affluent salinity (0.2–3.0 dS m^{-1}) (Table 4).

The correlation analysis between ΔCE and meteorological variables, although significant, showed low values for Pearson's correlation coefficient. The relative humidity of the air is inversely proportional to ΔCE and has a correlation coefficient ($p <$ 0.05) ranging from -0.29 to -0.72 for Tifton 85 grass and between -0.45 to -0.69 for vetiver grass. The combination of relative humidity and air temperature can influence the crop's evapotranspiration, resulting in a concentration of salts in HSSF-CW effluent located in hot and dry climate regions (Freedman *et al.* 2014). Solar radiation and wind speed act in parallel with the temperature in ΔEC since these variables are determinants for the magnitude of evapotranspiration.

Regarding the ET$_C$, there is a positive correlation for both species of grass, demonstrating that the loss of water in the system contributes to the increase in effluent EC, as suggested by Kadlec & Wallace (2009). It is important to note that the correlations were obtained for each level of affluent salinity, with the HSSF-CW mesocosm as the control volume and a period of 24 h to perform the water balance, unlike the previous topics, in which the comparison was performed between different productivities and affluent salinity levels. Vera *et al.* (2016) observed a positive correlation between ET$_C$ and the increase in EC effluents from HSSF-CW when the evapotranspired water volume exceeded 23% of the system affluent

Table 4 | Pearson's correlation analysis between ΔEC and meteorological variables monitored over the experimental period

Species	Variables	ET$_C$	Average air temperaure	Relative air humidity	Solar radiation	Wind speed
Tifton 85	ΔCE_0.2	0.14	0.10	−0.29	0.12	0.07
	ΔCE_0.5	0.37	0.05	−0.47	0.29	0.23
	ΔCE_1.0	0.06	−0.16	−0.31	0.20	0.11
	ΔCE_1.5	0.27	0.04	−0.48	0.36	0.23
	ΔCE_2.0	0.45	0.09	−0.54	0.38	0.25
	ΔCE_3.0	0.66	0.23	−0.72	0.53	0.40
Vetiver	ΔCE_0.2	0.33	0.15	−0.45	0.24	0.16
	ΔCE_0.5	0.39	0.10	−0.59	0.37	0.25
	ΔCE_1.0	0.46	0.09	−0.69	0.48	0.31
	ΔCE_1.5	0.58	0.28	−0.69	0.52	0.37
	ΔCE_2.0	0.54	0.23	−0.67	0.51	0.38
	ΔCE_3.0	0.58	0.24	−0.68	0.52	0.39

volume. For lower values, there was no change in the effluent EC. According to Fountoulakis *et al.* (2017), ET_C is the main factor responsible for the variation in EC in CWs, limiting the efficiency of removing salts from wastewater treated in these systems (Nan *et al.* 2020).

According to the data presented in Table 4, it is observed that, in general, there was an increase in Pearson's correlation coefficient concerning the levels of affluent salinity, confirming that the plant's growth maximize the effect of the meteorological variables in EC variation of the solution under treatment. Freedman *et al.* (2014) found an increase in effluent EC of HSSF-CW grown with *B. indica* associated with this plant's greater vegetative development. According to the authors, the maximum value of effluent EC coincided with the maximum plant growth period. Additionally, after cutting, there was an instant reduction in the effluent EC values.

Table 5 shows the correlation analysis between the EC variations in the solution treated in HSSF-CW for different affluent salinity levels.

Higher Pearson correlation coefficients are found among HSSF-CW solutions with similar salinity levels. As the comparison is made between more distant levels, such as 0.2 and 3.0 dS m^{-1}, the lower the correlation. This makes it evident that the concentration of ions in the solution under treatment and, consequently, the plants' vegetative development has a high influence on ΔEC.

When correlating ΔEC between grasses, there is a tendency to increase Pearson's correlation coefficient with increased salinity levels. These results corroborate the biomass productivity and ET_C as the main factors responsible for the EC's variation in the saturated solution of the HSSF-CW porous medium.

Water use efficiency in HSSF-CW

As discussed throughout this work, the relationship between aerial biomass productivity and the evapotranspiration of the plants grown in HSSF-CW is decisive for the increase or reduction of effluent EC. Therefore, in these cases, WUE can be used as a subsidy for decision making regarding the choice of species to be grown in the beds of CWs (Headley *et al.* 2012; Tuttolomondo *et al.* 2016), mainly when the effluent is used for agricultural fertigation in places with water deficit.

According to Figure 5, the salinity of the nutrient solution has little influence on the WUE of the Vetiver grass, however, there is a significant increase regarding the Tifton 85 grass. These results are related to the productivity of aerial biomass of grasses since salinity levels are associated with the availability of nutrients for crops. Initially, it is observed that the grasses showed similar biomass productivity and similar ET_C values (Figure 3). However, for higher salinity levels, Tifton 85 grass productivity increased disproportionately to the ET_C, resulting in high water use efficiency (Figure 5).

Higher efficiency in the use of water and, consequently, higher ion absorption by plants have the effect of reducing ΔEC in HSSF-CW. According to Tuttolomondo *et al.* (2016), water use efficiency can be used as a criterion for selecting species that have high phytoremediation potential, contributing to improving the efficiency of wastewater treatment in HSSF-CW. According to the same authors, if the CW was designed to ensure high effluent flow, plants with low ET_C and low WUE should be selected. However, if treatment efficiency and plant aerial biomass produced are important factors, plants' selection with the high WUE and low ET_C is recommended.

Table 5 | Pearson's correlation analysis between ΔEC for the different affluent salinity levels to HSSF-CW

Species	Variable	ΔCE_0.2	ΔCE_0.5	ΔCE_1.0	ΔCE_1.5	ΔCE_2.0	ΔCE_3.0
Tifton 85	ΔCE_0.2	1.00					
	ΔCE_0.5	0.70	1.00				
	ΔCE_1.0	0.26	0.61	1.00			
	ΔCE_1.5	0.25	0.63	0.89	1.00		
	ΔCE_2.0	0.27	0.60	0.81	0.95	1.00	
	ΔCE_3.0	0.36	0.67	0.60	0.82	0.90	1.00
Vetiver	ΔCE_0.2	1.00					
	ΔCE_0.5	0.85	1.00				
	ΔCE_1.0	0.67	0.90	1.00			
	ΔCE_1.5	0.48	0.77	0.89	1.00		
	ΔCE_2.0	0.37	0.70	0.85	0.96	1.00	
	ΔCE_3.0	0.35	0.69	0.84	0.96	0.98	1.00

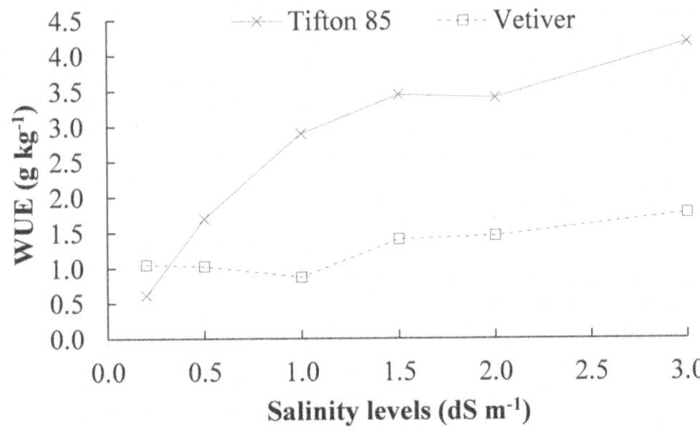

Figure 5 | Water use efficiency (WUE) of Vetiver and Tifton 85 grasses as a function of the affluent salinity levels from the HSSF-CW.

Figure 6 | Development of Tifton 85 (a, b) and Vetiver grasses (c, d) submitted to the affluent salinity levels of 0.2 dS m^{-1} (a, c) and 3.0 dS m^{-1} (b, d).

Thus, the WUE is considered an essential metric in selecting species most suitable for cultivation in HSSF-CW located in regions of an arid climate, where the excessive loss of water to the atmosphere is undesirable (Headley *et al.* 2012).

Effects of salinity on plant growth

The low affluent salinity levels affected the growth and development of the Vetiver and Tifton 85 grasses, resulting in lower biomass productivity. This result is associated with the availability of nutrients for crops, and consequently their uptake by the plants.

Tifton 85 grass showed a higher vegetative development compared to Vetiver grass when submitted to the highest affluent salinity levels. However, for low levels, Vetiver development was higher (Figure 6).

Considering the growth and development of the Vetiver in the nutrient solution, your cultivation is not considered suitable in HSSF-CW used for treatment of wastewater containing high nutrient concentrations. In relation to low concentration wastewater, Vetiver grass can be considered an ideal plant, since its growth and development is higher than that of Tifton 85 for low affluent salinity levels (Teixeira *et al.* 2021).

CONCLUSION

The Vetiver and Tifton 85 grasses' cultivation increased the effluent EC of the HSSF-CW, mainly for the lowest levels of affluent salinity.

The crops evapotranspiration contributes to an increase in effluent EC of the HSSF-CW. However, plant biomass productivity has a great influence on the ΔEC, so that the increase or reduction in EC of the treated wastewater depends on the relative magnitude of the transpiration processes and ion absorption by the plants.

The cultivation of different plant species in different climatic conditions, as well as in diverse wastewater compositions, results in different behaviors concerning the ΔEC of the solution treated in HSSF-CW. However, plants with low ET$_C$ and high plant biomass productivity are recommended for cultivation in HSSF-CW located in hot-dry climate regions to enable the reuse of effluents in agricultural fertigation.

ACKNOWLEDGEMENTS

The authors wish to thank CNPq for granting scholarships for research accomplishment.

DATA AVAILABILITY STATEMENT

All relevant data are included in the paper or its Supplementary Information.

REFERENCES

Almeida, A., Carvalho, F., Imaginário, M. J., Castanheira, I., Prazeres, A. R. & Ribeiro, C. 2017 Nitrate removal in vertical flow constructed wetland planted with Vetiveria zizanioides: effect of hydraulic load. *Ecological Engineering* **99**, 535–542.

Araujo, E. D., Borges, A. C., Dias, N. M. & Ribeiro, D. M. 2018 Effects of gibberellic acid on Tifton 85 bermudagrass (Cynodon spp.) in constructed wetland systems. *PloS one* **13** (10), 1–26.

Ayers, R. S. & Westcot, D. W. 1985 Water quality for agriculture. *FAO Irrigation and Drainage Paper 29 (Rev. 1)*. Food and Agriculture Organization of the United Nations, Rome, Italy, p. 186.

Bachand, P. A. M., Bachand, S., Fleck, J., Anderson, F. & Windham-Myers, L. 2014 Differentiating transpiration from evaporation in seasonal agricultural wetlands and the link to advective fluxes in the root zone. *Science of the Total Environment* **484**, 232–248.

Fia, R., Boas, R. B. V., Campos, A. T., Fia, F. R. & Souza, E. G. D. 2014 Removal of nitrogen, phosphorus, copper and zinc from swine breeding waste water by bermudagrass and cattail in constructed wetland systems. *Engenharia Agrícola* **34** (1), 112–113.

Fia, F. R. L., Matos, A. T., Fia, R., Borges, A. C. & Cecon, P. R. 2017 Effect of vegetation on flooded systems built to treat swine wastewater. *Engenharia Sanitária E Ambiental* **22** (2), 1–9.

Fountoulakis, M. S., Daskalakis, G., Papadaki, A., Kalogerakis, N. & Manios, T. 2017 Use of halophytes in pilot-scale horizontal flow constructed wetland treating domestic wastewater. *Environmental Science and Pollution Research* **24**, 16682–16689.

Freedman, A., Gross, A., Shelef, O., Rachmilevitch, S. & Arnon, S. 2014 Salt uptake and evapotranspiration under arid conditions in horizontal subsurface flow constructed wetland planted with halophytes. *Ecological Engineering* **70**, 282–286.

Gao, J., Wang, W., Guo, X., Zhu, S., Chen, S. & Zhang, R. 2014 Nutrient removal capability and growth characteristics of Iris sibirica in subsurface vertical flow constructed wetlands in winter. *Ecological Engineering* **70**, 351–361.

Headley, T. R., Davison, L., Huett, D. O. & Müller, R. 2012 Evapotranspiration from subsurface horizontal flow wetlands planted with Phragmites australis in sub-tropical Australia. *Water Research* **46**, 345–354.

Jesus, F. L. F., Matos, A. T. & Matos, M. P. 2020 Efficiency of horizontal subsurface flow constructed wetlands cultivated with grasses of different root systems. *Water Supply* **20** (8), 3318–3329.

Kadlec, R. H. & Wallace, R. D. 2009 *Treatment Wetlands*, 2nd edn. CRC Press, Boca Raton, FL, USA, p. 1016.

Lavrnié, S., Zapater-Pereyra, M. & Mancini, M. L. 2017 Water scarcity and wastewater reuse standards in Southern Europe: focus on agriculture. *Water, Air, & Soil Pollution* **228** (7), 228–251.

Liang, Y., Zhu, H., Bañuelos, G., Yan, B., Zhou, Q., Yu, X. & Cheng, X. 2017 Constructed wetlands for saline wastewater treatment: a review. *Ecological Engineering* **98**, 275–285.

Matos, A. T. 2015 *Solid Waste and Wastewater Analysis Manual*. Editora UFV, Viçosa, Brazil, p. 149.

Nan, X., Lavrnić, S. & Toscano, A. 2020 Potential of constructed wetland treatment systems for agricultural wastewater reuse under the EU framework. *Journal of Environmental Management* **275**, 111219.

Saraiva, C. B., Matos, A. T., Matos, M. P. & Miranda, S. T. 2018 Influence of substrate and species arrangement of cultivated grasses on the efficiency of horizontal subsurface flow constructed wetlands. *Engenharia Agrícola* **38** (3), 417–425.

Saraiva, C. B., Matos, A. T. & Matos, M. P. 2019 Extraction capacity of grasses grown in constructed wetland systems using different arrangements and substrates. *Engenharia Agrícola* **39** (5), 668–675.

Shelef, O., Gross, A. & Rachmilevitch, S. 2012 The use of Bassia indica for salt phytoremediation in constructed wetlands. *Water Research* **46** (13), 3967–3976.

Silva, M. S. G. M. E., Losekann, M. E. & Roston, D. M. 2017 *Evaluation of A Cultivated bed System with Recirculation for Fish Farming*. Embrapa Meio Ambiente, Jaguariúna, Brazil, p. 27.

Taiz, L., Zeiger, E., Moller, I. M. & Murphy, A. 2017 *Plant Physiology and Development*, 6th ed. Artmed, Porto Alegre, Brazil, p. 888.

Teixeira, D. L., Matos, A. T., Matos, M. P., Miranda, S. T. & Teixeira, D. V. 2021 Modeling of productivity and nutrient extraction by the Vetiver and Tifton 85 grasses grown in horizontal subsurface flow constructed wetlands. *Journal of Environmental Science and Health, Part A* **56** (3), 248–256.

Tuttolomondo, T., Licata, M., Leto, C., Leone, R. & La Bella, S. 2015 Effect of plant species on water balance in a pilot-scale horizontal subsurface flow constructed wetland planted with *Arundo donax L.* and *Cyperus alternifolius L.* – Two-year tests in a Mediterranean environment in the West of Sicily (Italy). *Ecological Engineering* **74**, 79–92.

Tuttolomondo, T., Leto, C., La Bella, S., Leone, R., Virga, G. & Licata, M. 2016 Water balance and pollutant removal efficiency when considering evapotranspiration in a pilot-scale horizontal subsurface flow constructed wetland in Western Sicily (Italy). *Ecological Engineering* **87**, 295–304.

Vera, I., Verdejo, N., Chávez, W., Jorquera, C. & Olave, J. 2016 Influence of hydraulic retention time and plant species on performance of mesocosm subsurface constructed wetlands during municipal wastewater treatment in super-arid areas. *Journal of Environmental Science and Health, Part A* **51**, 105–113.

First received 23 November 2021; accepted in revised form 17 February 2022. Available online 28 February 2022

© 2022 The Authors

doi: 10.2166/wst.2022.083

Performance intensification of constructed wetland technology: a sustainable solution for treatment of high-strength industrial wastewater

N. Nurmahomed [a],*, A. K. Ragen [a] and C. M. Sheridan [b]

[a] Chemical and Environmental Engineering department, Faculty of Engineering, University of Mauritius, Reduit, Mauritius
[b] Centre in Water Research and Development, School of Geography, Archaeology and Environmental Studies, University of the Witwatersrand, Johannesburg, South Africa
*Corresponding author. E-mail: nazeemah.nurmahomed@gmail.com

NN, 0000-0003-0802-1036; AKR, 0000-0001-7561-1816; CMS, 0000-0002-5913-3428

ABSTRACT

The objectives of this study were to: (1) assess the intensification of chemical oxygen demand (COD) and phosphate (PO_4-P) removal; and (2) generate a set of rate constants of COD degradation (k_{COD}) and phosphate (k_{PO4-P}) removal for the treatment of industrial wastewater (WW) using intensified adsorption beds. Two horizontal subsurface flow constructed wetlands (HSSFCWs) filled with coal ash and alum sludge and two conventional HSSFCWs packed with gravels were operated with different loadings of COD and PO_4-P at a hydraulic retention time (HRT) of 24 hrs at water depth of 0.40 m. The bed performance was analysed for COD and PO_4-P removal efficiency. The intensified HSSFCWs outperformed the control beds by a mean COD and PO_4-P removal efficiency of 43 and 49%, respectively. The progression of COD and PO_4-P removal along the system was fitted into the first-order plug flow model (K-C model). In this study the k_{COD} values ranged from 0.36 to 0.65 m/d with a mean of 0.46 ± 0.08 m/d ($n = 30$). The k_{PO4-P} values ranged from 0.74 to 1.76 m/d and averaged to 1.23 ± 0.37 m/d ($n = 30$), irrespective of the condition applied. Hence, these data can be used for future projects using HSSFCWs to treat industrial wastewater.

Key words: adsorption, constructed wetland, intensified, K-C model

HIGHLIGHTS

- Industrial wastewaters are strong organic sources of pollution.
- Industrial effluents are not commonly treated using conventional HSSFCW.
- Addition of an adsorption mechanism using coal ash and alum sludge in the system intensifies bed performance.
- Such intensified HSSFCWs are a sustainable and low-cost treatment system.
- Rate constants are important design parameters to determine the size of a HSSFCW.

GRAPHICAL ABSTRACT

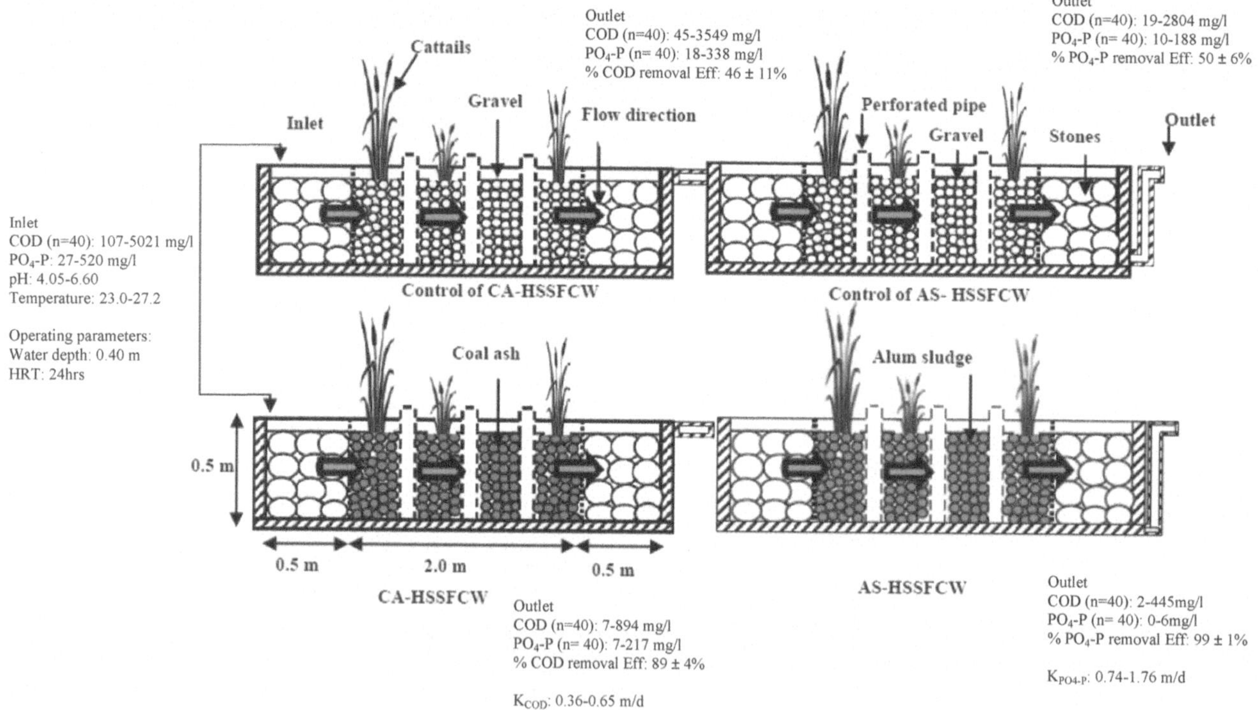

INTRODUCTION

The discharge of untreated industrial effluents is considered as one of the main sources of water pollution that impact negatively on the environment and human health (Swain *et al.* 2018). Hence, purification of such effluents is necessary, but this may be complicated due to the presence of complex chemical profiles (Yang *et al.* 2018). Constructed wetland (CW), also known as the 'natural biological reactor', is a sustainable alternative for the purification of industrial wastewater (WW) (Khan *et al.* 2009). This nature-based biological system comprises macrophages, substrates and microbial communities and treats wastewater through biological, chemical and physical processes (Saeed & Khan 2019). CWs are effective in contaminant removal, require simple maintenance and do not typically require the input of energy and/or chemicals (Zhuang *et al.* 2019).

Among the various types of constructed wetlands that exist, horizontal subsurface flow CWs (HSSFCWs), the focus of this paper, have obtained much attention globally (Zhou *et al.* 2019). Wu *et al.* (2015) reported that the direct input of effluent having high loads of organic matter (OM) in the system can affect the growth and health of the plants as well as the treatment performance. Rajkumar *et al.* (2010) stated that to implement a conventional biological treatment system, the biological oxygen demand/chemical oxygen demand (BOD/COD) ratio, which signifies the biodegradability of WW, should be greater than 0.6. Effluents having a BOD/COD ratio less than 0.6 cannot be treated effectively by biological purification system alone; additional treatment is typically required.

CWs have developed from passive systems to advance engineered systems, especially those known as intensified wetlands. These have the capacity of degrading contaminants 10–1,000 times faster than conventional CW (Nivala *et al.* 2019). Intensified CWs can be designed differently and built into special configurations to increase treatment performance (Wu *et al.* 2014). The intensified systems provide advantages, such as they require relatively small space to treat WW with an elevated level of pollutants. Among the intensified systems that exist, the integration of adsorption removal mechanisms into CW is believed to be a useful approach because of various benefits such as simple operation, fast elimination of pollutants, low cost and prevention of production of secondary contaminants (Nivala *et al.* 2019).

The substrate is an important component of a CW system especially for subsurface CWs since it allows attachment of microorganisms, serves as a growing platform for the macrophages and has the potential to eliminate pollutants (Wu *et al.* 2015). Yang *et al.* (2018) discuss that the substrate can either be natural (such as gravel) or artificial (such as by-products from industrial processes). Wahyuni *et al.* (2018) found that alum sludge (AS) is a good adsorbent of phosphate (PO_4-P) and coal ash is a potent adsorbent for the removal of COD. However, there has been no study on the integration of coal ash and AS into HSSFCWs purifying high-strength industrial effluents. Also, there is little published data on the COD and PO_4-P removal rate constants (k_{COD} and k_{PO4-P}) for this type of technology treating effluent with elevated levels of COD and PO_4-P. Inaccurate data for k_{COD} and k_{PO4-P} can lead to over- or underdesign and ultimately can cause poor performance and/or failure of the HSSFCW.

Hence, the main aims of this study are: (1) to investigate the intensification of COD and PO_4-P removal in a coal bottom ash (CBA) and AS integrated HSSFCW; and (2) to develop a range of rate constants of COD/ PO_4-P removal (k_{COD} and k_{PO4-P}) and consequent operational criteria for such system.

METHODOLOGY

Batch experiment

Batch analysis was performed as per the method of Shah *et al.* (2013) to obtain the adsorptive capacity and adsorptive index of coal ash and AS. For the investigation of COD adsorption by activated CBA, a molasses solution having a concentration of 3,435 mg/l of COD was prepared. A phosphate solution of concentration 610 mg/l of PO_4-P was prepared to assess phosphate adsorption by AS.

Batch experiments were performed with sets of beakers comprising 100 ml molasses solution in contact with of 2 g, 4 g, 6 g, 8 g, 10 g and 12 g of activated CBA. The beakers were shaken continuously for 6 hours at 160 rpm until equilibrium was obtained. Samples were taken at an interval of two hours over a period of six hours. A Batch test for the adsorption of phosphate by AS was performed using 0.5 g, 1.0 g, 1.5 g, 2.0 g, 2.5 g and 3.0 g of AS in contact with 100 ml of PO_4-P solution. The samples were then analysed for COD and PO_4-P. The determination of COD was carried out according to APHA (2005) Standard Method of Examination of Water and Wastewater. The USEPA approved Hach DR-2500 light spectrophotometer was utilised to determine PO_4-P by the PhosphoVer3 (ascorbic acid) method.

Continuous column analysis

Continuous column analysis was carried out according to the method described by Wahyuni *et al.* (2018) to obtain the saturation time of the activated CBA and AS. For COD adsorption, 20 g of CBA was placed in a 5 cm diameter glass column and for phosphate adsorption; 15 g of AS was used. Molasses and phosphate solution were fed from the top of the column at a flowrate of 3 ml/min. Samples from the bottom were extracted each 15 minutes over a period of three hours and were then analysed for COD and PO_4-P.

Adsorption isotherms

For better determination of the behaviours, interactions and adsorptive capacity of CBA and AS, Freundlich (Equation (1)), Langmuir (Equation (2)) and Dubinin-Radushkevich (Equation (4)) sorption isotherms were used.

$$ln q_e = ln k_F + \frac{l}{n} ln C_e \tag{1}$$

where q_e is the equilibrium adsorption capacity of adsorbent in mg/g; k_F stands for the Freundlich constant, adsorption capacity in mg/g; n is the Freundlich adsorption intensity, which is dimensionless and it represents favourable adsorption if n-value lies between 0 and 10 (Mirzaei & Javanbakht 2019)

$$\frac{C_e}{q_e} = \frac{C_e}{Q_{max}} + \frac{1}{Q_{max}k_l} \tag{2}$$

C_e denotes the equilibrium concentration in mg/l; Q_{max} is the maximum adsorption capacity of adsorbents in mg/g; k_L is the Langmuir constant in L/mg. The favourability of adsorption can be determined by R_L, which is a dimensionless separation factor as shown in Equation (3). Favourable adsorption occurs when R_L lies between 0 and 1, linear when R_L is equal to 1 and irreversible when R_L is equal to 0 (Gebreegziabher *et al.* 2019).

$$R_L = \frac{1}{1 + bC_o} \tag{3}$$

$$\ln q_e = \ln q_{max} - \beta\varepsilon^2 \tag{4}$$

In Equation (4), β represents the mean adsorption energy in mol^2/kJ^2 and ε symbolises the Polanyi potential, which is calculated using Equation (5).

$$\varepsilon = RT \ln \left(1 + \frac{1}{C_e}\right) \tag{5}$$

R represents the universal gas constant, which is 8.3145 J/mol/K and T is the absolute temperature in Kelvin (°K). The Dubinin-Radushkevich isotherm was utilised to find out the type of sorption behaviour (Tolić *et al.* 2019). The mean free energy of sorption, denoted by E (kJ mol^{-1}), is associated to the reaction mechanism. A value less than 8 kJ/mol represents physical sorption, while those that lies between 8 and 16 kJ mol^{-1}, indicates chemical sorption. Those greater than 16 kJ mol^{-1} signify a very strong chemical sorption (Yan *et al.* 2014). The mean sorption energy can be calculated by Equation (6).

$$E = \frac{1}{\sqrt{2\beta}} \tag{6}$$

Yoon-Nelson model

Among the models related for the continuous column experiment, the Yoon Nelson model as shown in Equation (7) was found to be a less complex model to determine the saturation time of adsorbents since it does not need any information relating the properties of the adsorbent or adsorbate (Jang & Lee 2019).

$$ln\frac{C_t}{C_o - C_t} = k_{YN}t - \tau k_{YN} \tag{7}$$

In Equation (7), k_{YN} represents the rate constant in min^{-1} and τ stands for the time required for 50% adsorbate breakthrough (in minutes).

Experimental set-up of coal ash and alum sludge HSSFCW

For the experiment, four HSSFCWs were constructed at Reduit, University of Mauritius (20°13′59.9″S 57°29′57.5″E). Two parallel treatment systems were set up whereby each was established with two HSSFCWs in sequence, each having dimensions of 3.0 m × 0.5 m × 0.5 m. One train, denoted as 'intensified beds', comprised of one HSSFCW in which activated coal ash was used as substrate (denoted as CA-HSSFCW), followed by the second train, which was packed with AS (denoted as AS-HSSFCW). Control beds were composed of two conventional HSSFCWs in series. For both experiments, the train was packed with gravel (denoted as CCA-HSSFCW for the coal ash and as CAS-HSSFCW for the AS). The inlet and outlet structures were comprised of stones with a size of 50–200 mm. Cattails (*Typha latifolia*) were planted in all the four beds at a planting density of 8 plants/m^2. Synthetic industrial wastewater simulating COD concentrations ranging from 107 to 5,021 mg/l was prepared by diluting the required amount of molasses in a feed tank having a capacity of 2 m^3. The required amount of potassium dihydrogen orthophosphate and potassium nitrate (nutrients for macrophage growth) to achieve PO$_4$-P concentration of 27–520 mg/l were diluted in the same feed.

As this study is based on the purification of high-strength effluent, molasses were utilised to imitate a real industrial WW since it consists of hard fractions as well as organic components (Meng *et al.* 2017). Synthetic wastewater was prepared two or three times per week. Furthermore, three perforated pipes were placed within each bed to take samples as well as to allow oxygen to infiltrate the substrate. A ball valve was fitted in the inlet line to control the flowrate of the WW fed to the HSSFCW. The experimental set up is illustrated in Figure 1.

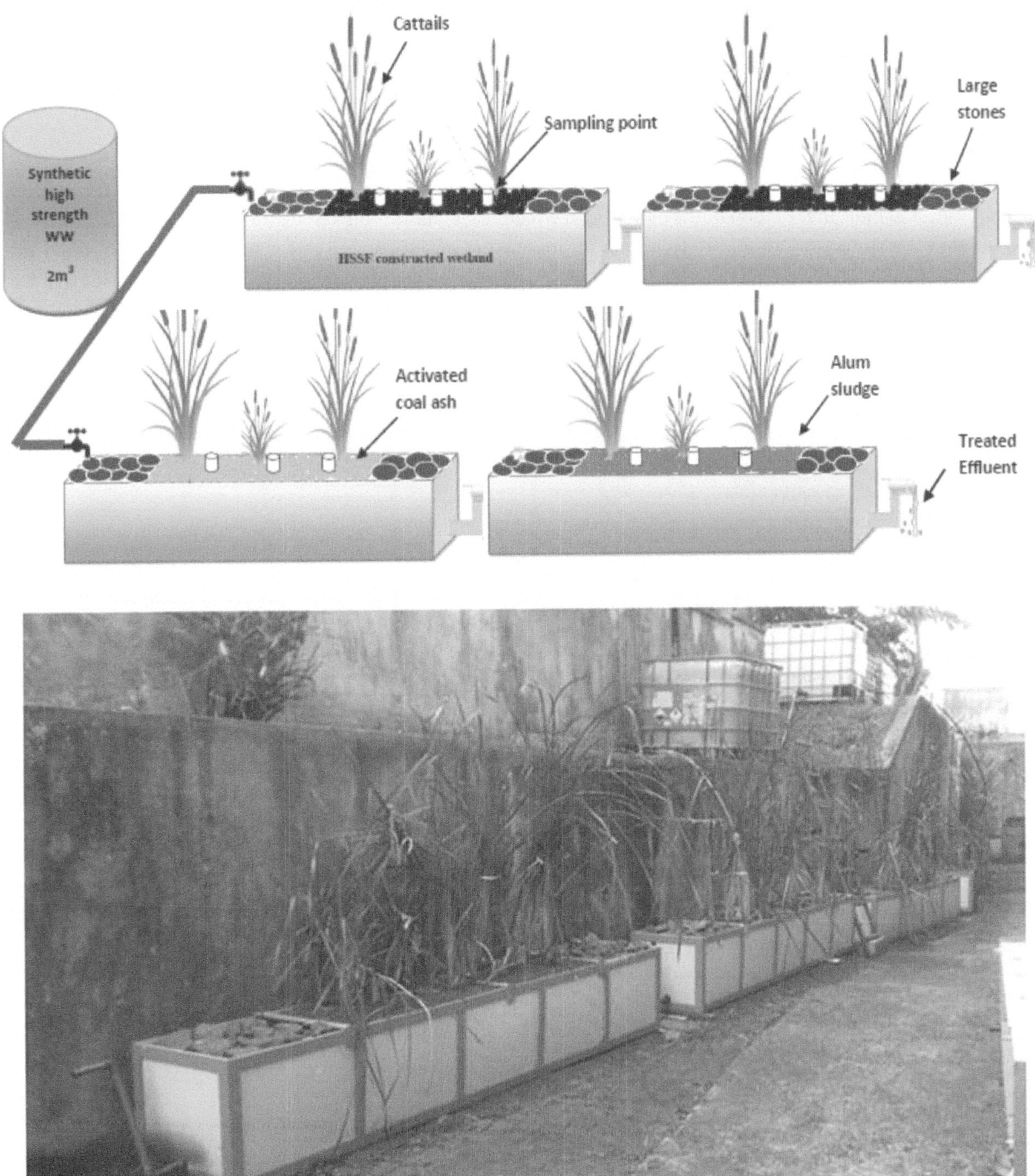

Figure 1 | Intensified and conventional experimental HSSFCWs.

Operating the pilot system

The system was started up for a continuous period of ten weeks with synthetic wastewater of low COD loading (107–130 mg/l) and phosphate loading (23–35 mg/l PO_4-P) at retention time of 24 hrs and a water depth of 0.40 m. After successful start-up, the beds were operated at medium organic loads (912–1,563 mg/l) and phosphate loadings (316–380 mg/l PO_4-P) for a continuous period of 18 weeks. Thereafter, the experimental beds were run at a higher COD loads of range 4,801–5,021 mg/l and phosphate concentrations of range 488–520 mg/l PO_4-P for 12 weeks. Samples were taken from the inlet port, the three intermediate sampling ports and outlet pipes, and were analysed for COD and PO_4-P. The inlet, three intermediates and outlet sampling ports (denoted respectively as S_1, S_2, S_3, S_4 and S_5) are shown in Figure 2. The samples were also monitored for pH and temperature.- The pH was monitored by means of a digital pH meter (model: EUTECH Instrument cyber scan pH 11). A thermometer was used to measure the temperature of the samples.

Kinetics and first-order plug flow modelling

Kinetics for a chemical engineer are considered as an essential element in designing at a professional stage (Sheridan *et al.* 2014).With respect to time, the reaction rate is the speed at which reactants are converted into products. The first-order rate of reaction is proportional to concentration. According to Sheridan *et al.* (2014), the linear reaction rate (first-order rate) is usually used for CW unless they operate beyond saturation conditions. Microbial degradation along the HSSFCW follows a K-C model, commonly known as a first-order plug flow model, from which the Kickuth equation (Equation (8)) was derived.

$$A = \frac{Q(\ln C_o - \ln C_t)}{k_{COD}} \tag{8}$$

where A: bed surface area, m^2; Q: flow rate of effluent, m^3/day; C_o: inlet COD concentration, mg/L; C_t: outlet COD concentration, mg/L and k_{COD}: COD rate constant, m/day.

The first-order plug flow model can be rewritten as:

$$C_t = C_0 e^{-kt} \tag{9}$$

where C_t: effluent concentration of COD/PO_4-P (mg/l); C_o: inlet concentration of COD/PO_4-P (mg/l); k: rate constant, (/d); t: time of degradation (d).

The K-C model fits best for an ideal flow condition and also shows that the exponential decrease of pollutants tends to a zero value. Von Sperling & de Paoli (2013) showed that a HSSFCW does not behave as an ideal flow reactor and an outlet concentration of zero was never found. Therefore, the K-C* model is recognised as a more appropriate model than the K-C model to be used in a HSSFCW since a non -zero outlet concentration is mostly found. Equation (10) shows the K-C* model.

$$C_t - C^* = (C_0 - C^*)e^{-kt} \tag{10}$$

The rate constants k_{COD} and $k_{PO4\text{-}P}$ are important parameters for the sizing of a HSSFCW. This study attempts to propose a range of these rate constants for an intensified bed purifying high-strength industrial WW utilising a first-order plug-flow

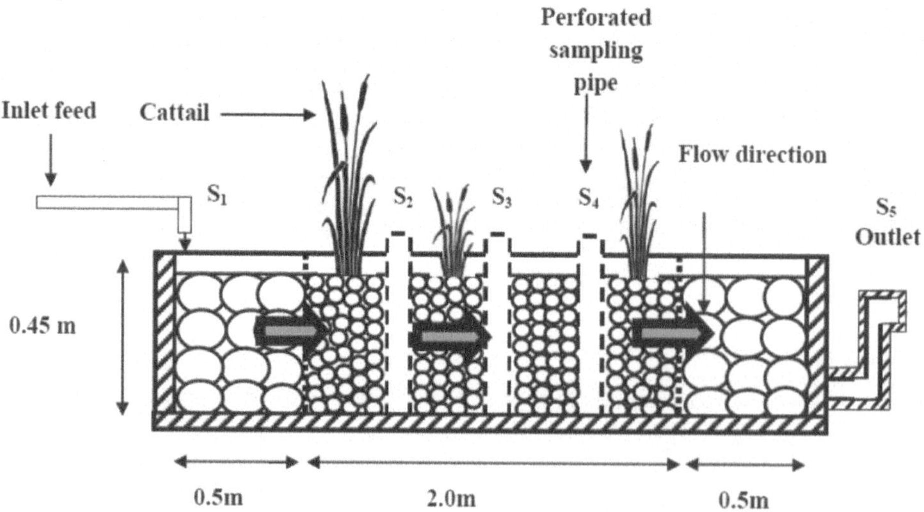

Figure 2 | HSSFCW showing sampling ports.

model. These can be calculated using Equation (11).

$$k_{(COD or PO4-P)} = k_T dn \tag{11}$$

where d is the depth of water (m) and n: porosity of the substrate (dimensionless).

Operating parameters such as hydraulic retention time (HRT), hydraulic loading rate (HLR) and organic surface loading rate (OSLR) affect the efficiency of the system generally. These are represented by Equations (12)–(14), respectively.

$$HRT = \frac{Lwdn}{Q} \tag{12}$$

where HRT = hydraulic retention time; L = length of bed, m; w = width of bed, m; d = average depth of bed, m; n = porosity; %; Q = flow, m³/d;

$$HLR = \frac{Q}{A_h} \tag{13}$$

where HLR = hydraulic loading rate, m/d; Q = flow rate, m³/d; A_h = surface area of bed, m²

$$OSLR = \frac{QxS_o}{A_h} \tag{14}$$

where $OSLR$ = organic surface loading rate, kgCOD/m².d; Q = flow rate, m³/d; S_o = inlet COD, kg/m³; A_h = surface area of bed, m².

Statistical analysis

Standard deviations and mean values of: (1) inlet COD and PO_4-P; (2) outlet COD and PO_4-P; (3) OSLR and phosphate loading rate (PLR); (4) efficiency of COD and PO_4-P removal; and (5) k_{COD} and k_{PO4-P} were determined using Excel™ software. IBM SPSS Software was used to perform: (1) paired-sample T-test to determine if the mean difference of the characteristic of the inlet and outlet were statistically significant at 95% confidence level. If the characteristic of the inlet and outlet mean are alike, then the null hypothesis tested is not discarded; and (2) Pearson correlation test to assess the statistical significance of the operating parameters such as OSLR and PLR on bed performance. Furthermore, the observed COD and PO_4-P data obtained along each HSSFCW were used in the K-C and K-C* models to confirm the results of this study. The outcomes should fit these two models as the beds function under a plug-flow condition (Von Sperling & de Paoli 2013).

RESULTS AND DISCUSSIONS

Adsorption capacity & time saturation

Table 1 shows the results obtained from the batch analysis using activated CBA and AS.

From Table 1, the adsorption intensity (n value) for AS and coal ash lies between 1 and 10 representing favourable adsorption. Similar results were obtained for Langmuir model where R_L lies between 0 and 1 indicating that coal ash and AS are good adsorbents. The Dubinin-Radushkevich model showed there to be physical sorption since the mean sorption was less than 8 kJ/mol. Table 2 shows the results obtained from the continuous column analysis, which was substituted into the Yoon Nelson Model to determine the saturation time of the coal ash and AS.

From Table 2, a better fit was obtained for Yoon Nelson model (high R^2-values) for both AS and coal ash (CA). AS required 12 minutes per gram to be fully saturated. AS adsorbed phosphate more rapidly than COD showing its higher preference for phosphate. The same explanation can be made for COD removal using coal ash.

Characterization of influent and effluent

The characteristics of the inlet and outlet effluent of the four HSSFCWs are given in Tables 3 and 4, respectively.

Table 1 | Adsorption isotherm parameters for COD and phosphate adsorption onto CBA and AS[a]

Adsorbent	Adsorbate	Freundlich isotherm			Langmuir isotherm				Dubinin-Radushkevich isotherm			
		n	k_f	R^2	Qmax	$K_l \times 10^{-3}$	R_L	R^2	B	Qmax	E	R^2
Alum sludge	PO_4-P	2.26	5.93	0.97	67	24.0	0.10	0.98	0.00003	39	129	0.71
	COD	2.18	3.52	0.75	143	0.1	0.16	0.65	0.009	46	7	0.76
Coal ash	PO_4-P	1.89	0.98	0.82	28	13.0	0.10	0.97	0.00064	22.9	28	0.65
	COD	2.82	3.51	0.97	63	1.3	0.13	0.96	0.04	122	4	0.55

[a]n represents Freundlich adsorption intensity (dimensionless); k_f is the adsorption capacity (mg/g); Qmax is the equilibrium adsorption capacity of adsorbent (mg/g); K_l is the Langmuir constant related to the rate of adsorption (L/mg); R_L is the dimensionless equilibrium parameter; B stands for mean adsorption energy(mol^2/J^2) and E is the mean free energy sorption (J/mol).

Table 2 | Fixed bed adsorption column data and parameters obtained for COD and phosphate removal by using coal ash and AS at different conditions[a]

Adsorbent	Adsorbate	Co	Mass	v	k_{YN}	T	R^2
Alum sludge	PO_4-P	610	15	3	−0.030	12	0.848
	COD	1,684	15	3	−0.019	20	0.896
Coal ash	PO_4-P	610	20	3	−0.036	6	0.926
	COD	9,160	20	3	−0.037	9	0.882

[a]Yoon-Nelson parameters where C_o is the initial concentration (mg/L); v is the flowrate (ml/min); k_{YN} is the rate constant (L/min); T is the time saturation for one gram of adsorbate to get saturated (min).

Table 3 | Influent wastewater characteristics

	Influent Parameters							
	COD (mg/l)		PO_4-P (mg/l)		pH		Temp (°C)	
Strength of WW	Range	Mean ± SD	Range	Mean ± SD	Range	Mean ± SD	Range	Mean ± SD
Low ($n = 10$)	107–130	115 ± 7	27–71	58 ± 13	5.41–6.60	6.14 ± 0.4	23.2–25.2	24.2 ± 0.5
Medium ($n = 18$)	912–1,563	1,150 ± 172	316–380	351 ± 20	4.59–5.77	4.86 ± 0.30	23.2– 27.2	25.4 ± 1.2
Strong ($n = 12$)	4,801–5,021	4,987 ± 60	488–520	506 ± 10	4.05–4.45	4.18 ± 0.12	23–27.0	24.1 ± 1.2

Table 4 | Outlet wastewater characteristics for the intensified and control beds

		Intensified beds				Control Beds			
		CA-HSSCW		AS-HSSCW		CCA-HSSCW		CAS-HSSCW	
Parameters	N	Range	Mean ± SD	Range	Mean ± SD	Range	Mean ± SD	Range	Mean ± SD
COD (mg/l)	40	7–894	279 ± 322	2–445	130 ± 177	45–3,549	1,268 ± 1,429	19–2,804	761 ± 995
PO_4-P mg/l)	40	7–217	126 ± 69	0–6	1 ± 2	18–338	192 ± 97	10–188	95 ± 52
pH	40	5.89–6.99	6.67 ± 0.24	6.40–7.96	7.17 ± 0.4	6–6.9	6.41 ± 0.23	6.39–7.14	6.8 ± 0.19
Temp(°C)	40	22.9–27.1	24.7 ± 1.2	22.6–27.1	24.7 ± 1.2	22.8–27.1	24.7 ± 1.2	23–27.1	24.7 ± 1

The COD, PO_4-P, pH and temperature for the influent ranged from 107 to 5,021 mg/l, 27 to 520 mg/l, 4.05 to 6.60 and 23 to 27.2 °C and averaged to 2,042 ± 2,000 mg/l, 324 ± 170 mg/l, 5.03 ± 0.82, 24.7 ± 1.2 °C respectively. A strong influent having high COD and PO_4-P loadings was used since this study involved the purification of high-strength industrial effluent.

Table 4 shows the range, the mean and the standard deviation of the outlet wastewater for the intensified beds (CA-HSSFCW and AS-HSSFCW) and control beds (CCA-HSSFCW and CAS-HFSSCW) irrespective of the operating conditions applied. Greater values for the standard deviations were obtained for all the influent parameters in contrast with the effluents of the four HSSFCWs since three types of WW were prepared. In addition, the standard deviation for the outlet of the intensified and conventional beds was lower than the inlet showing that in all cases, regardless of remedial activity, the systems moderate the effluent load.

The *p*-value obtained for the means of the inlet and outlet COD and PO_4-P from the paired sample T-test using SPSS Software version 23 was less than 0.05, which means that there was a statistically significant difference between inlet & outlet COD and PO_4-P (at 95% confidence). The null hypothesis was hence rejected. This means that the synthetic industrial WW was treated by the four experimental beds in terms of COD and PO_4-P.

COD removal efficiency

In Figure 3, a comparison of the COD removal efficiency of the CA-HSSFCW and CCA-HSSFCW at a depth of 0.40 m for a retention time of 24 hours is shown.

The COD removal efficiency for the intensified coal ash HSSFCW and conventional HSSFCW ranged, respectively, from 82 to 96% and from 29 to 62% irrespective of the operating conditions that were applied. The mean of the COD removal efficiencies for the CA-HSSFCW and CCA-HSSFCW were 89 ± 4% and 46 ± 11% ($n = 40$), respectively. Therefore, it is concluded that an additional mean (or intensified) removal efficiency of COD of 43% has been accomplished by an intensified CA-HSSFCW bed purifying the high-strength synthetic industrial effluent. Furthermore, the four beds of this study have reached an established state as depicted in Figure 1. Figure 4 depicts the start-up of the systems and Figure 1 shows the bed with well-developed cattails plants.

Table 5 shows the COD removal efficiency for CA-HSSFCW and CCA-HSSFCW. There was an increase in the mean OSLR from 0.024 ± 0.001 ($n = 10$), 0.238 ± 0.035 ($n = 18$) and 1.030 ± 0.013 kgCOD/m^2d ($n = 12$) which represents a range of approximately 9–43 times the loading rate of COD.

Udom *et al.* (2018) discussed that there is a decline in the microbial activities, which leads to a decrease in the degradation of OM when the load increases. Similar results were observed in this study for the CA-HSSFCW and the CCA-HSSFCW (after excluding the start-up period of the system since all the four beds were not yet established at that time and the cattail plants were not well developed). Biodegradation is considered to be the most suitable process for the removal of OM in an effluent having elevated concentration of COD as stated by Stefanakis *et al.* (2016), However, as the OLR increases (as shown in Table 5) a significant decrease in the performance of the conventional system was observed compared to that of the

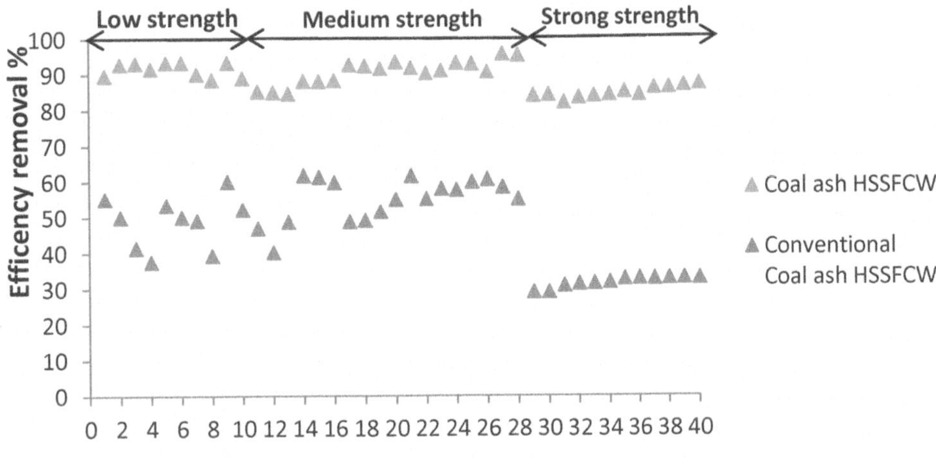

Figure 3 | Comparing COD removal efficiency for CA-HSSFCW and CCA-HSSFCW.

Figure 4 | The four beds (start-up phase in 2019).

Table 5 | Performance of CA-HSSFCW and CCA-HSSFCW at different strength of WW

| | | CA-HSSFCW | | | CCA-HSSFCW | | |
| | | OSLR (kgCOD/m²·d) | COD efficiency% | | OSLR (kgCOD/m²·d) | COD efficiency% | |
Parameters	N	Mean ± SD	Range	Mean ± SD	Mean ± SD	Range	Mean ± SD
Low	10	0.024 ± 0.001	88–93	91 ± 2	0.023 ± 0.01	38–60	49 ± 7
Medium	18	0.238 ± 0.035	85–96	91 ± 3	0.235 ± 0.035	40–62	55 ± 6
Strong	12	1.030 ± 0.013	82–87	85 ± 2	1.016 ± 0.013	29–33	32 ± 1

intensified beds. This provides evidence that the removal of the elevated amount of COD in the synthetic effluent was substrate dependent. When the OSLR increased from a mean of 0.238 ± 0.035 ($n = 18$) to 1.030 ± 0.013 ($n = 12$) kg COD/m²d, the COD removal efficiency dropped from a mean of $91 \pm 3\%$ ($n = 18$) to $85 \pm 2\%$ ($n = 12$), that is by an average of 6 percentage points whereas for the conventional HSSFCWs, when the OSLR increased from a mean of 0.235 ± 0.035 to 1.016 ± 0.013 kg COD/m²d, the removal efficiency of the COD decreased significantly from $55 \pm 6\%$ to $32 \pm 1\%$ that is by 23 percentage points. Mora-Orozco *et al.* (2018) studied a wetland system which was packed with a combination of soil, red volcanic rock and sand planted with *Typha* and *Scirpus* species. They obtained a removal efficiency of 86, 76 and 77% for treating a wastewater having a COD concentration of 400, 800 and 1,200 at an HRT of 10 days.

Since the operating factors HLR, HRT and depth of water remained unvaried during this study, the sole contributor for this intensified performance for the intensified systems was due to the mechanism of adsorption utilising coal ash as a medium. There was a slight decrease in the performance of the intensified beds compared to the conventional one because coal ash was used as media enhancing more removal of COD. It can be concluded that COD from the WW was removed not only by microbial degradation and plant uptake but by adsorption too. Hence, the adsorption mechanism contributes extensively to CW technology purifying high-strength industrial effluent.

Shepherd *et al.* (2001) demonstrated that 97% of COD is expected to be removed if the COD concentration of winery WW is less than 5,000 mg/l. However, when the organic matter concentration increased from 12,800 mg/l to 16,800 mg/l, the purification performance fell significantly, and the wetland plants found at the inlet structure of the system turned yellow and ultimately died. Zingelwa & Wooldridge (2009) studied the tolerance capacity of wetland plants which included *Typhalatifolia, Juncusacutus*, and *Scirpus maritimus*, which were involved in the treatment of winery WW containing elevated levels of organic matter. Their investigation revealed that the macrophytes were strong and healthy at a concentration of less than 5,000 mg/l and when the COD concentration reached 15,000 mg/l, the plants were stressed, yellowed and died. For this study, at a COD concentration ranging from 912 mg/to 1,563 mg/l, healthy plants were detected for both the intensified and conventional systems. Only some yellow shoots and leaves appeared in the control HSSFCW when the system was subjected to a WW having a strong concentration of COD of 4,801–5,021 mg/l. Nonetheless the plants from the CA-HSSFCW were not affected since adsorption mechanism was applied to the system. One more observation that has been made during the investigation was that the cattails grew and developed quickly compared to those from the conventional systems since most of the contaminants were adsorbed and hence, the cattails were under less metabolic stress. It was found that the performance of conventional beds decreased with increasing COD loadings. This indicates that conventional CWs were not able to treat effectively strong organic WWs. Thus, it may not be appropriate to apply CWs with gravel substrate to treat high-strength organic wastewater.

Phosphate removal efficiency

Figure 5 and Table 6 show the performance of the AS intensified systems (AS-HSSFCW) and the conventional AS systems (CAS-HSSFCW) in terms of phosphate removal efficiency at a retention time of 24 hrs and water depth of 0.40 m.

The phosphate removal efficiency of the AS-HSSFCW and CAS-HSSFCW, irrespective of the conditions applied ranged from 97 to 100% and 41 to 60%, respectively. The corresponding means for the phosphate removal efficiency for the AS-HSSFCW and CAS-HSSFCW were $99 \pm 1\%$ and $50 \pm 6\%$ ($n = 40$), respectively. It can therefore be concluded that a mean additional phosphate removal efficiency of 49% was achieved by an intensified CW purifying synthetic industrial wastewater. Since a low PO_4-P removal efficiency was obtained for the control beds packed with gravels, it is confirmed that gravel did not contribute substantially to the elimination of phosphate. Hence gravel is regarded as a poor adsorbent for phosphate removal (Wu *et al.* 2015).

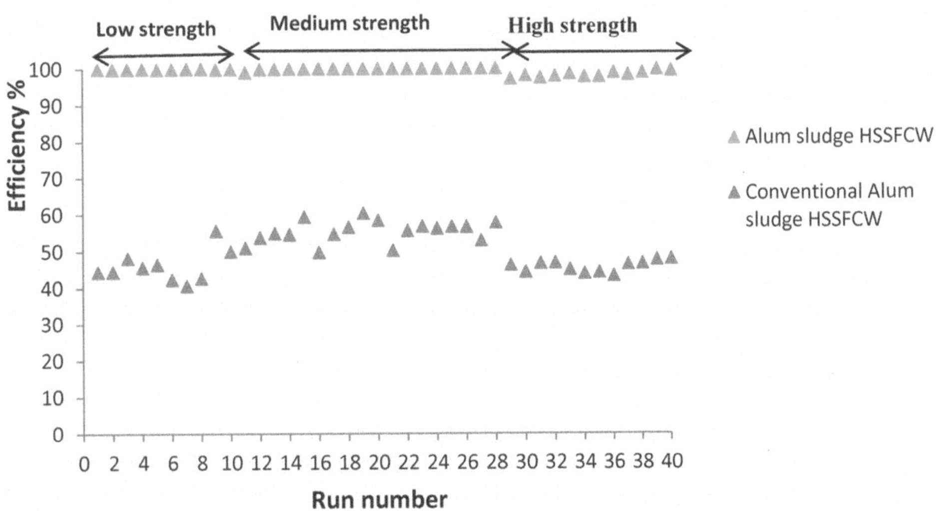

Figure 5 | Comparing PO_4-P removal efficiency for intensified AS-HSSFCW with CAS-HSSFCW.

Table 6 | Performance of AS and conventional HSSFCW at different strength of WW

| Parameters | n | AS-HSSFCW | | | CAS-HSSFCW | | |
| | | PLR (kg PO_4/m²d) | PO_4-P efficiency % | | PLR (kg PO_4/m²d) | PO_4-P efficiency% | |
		Mean ± SD	Range	Mean ± SD	Mean ± SD	Range	Mean ± SD
Low	10	0.0033 ± 0.0011	100–100	100 ± 0	0.0085 ± 0.0022	41–56	46 ± 4
Medium	18	0.030 ± 0.003	99–100	100 ± 0	0.043 ± 0.004	50–60	55 ± 3
Strong	12	0.039 ± 0.003	97–100	98 ± 1	0.060 ± 0.005	43–48	46 ± 2

Results for low strength WW were excluded since it was the start-up period of the systems where the HSSFCWs were not stable, and the cattails were not yet developed. The mean PLR in the CAS-HSSFCW increased from 0.043 ± 0.004 ($n = 18$) to 0.060 ± 0.005 kg PO_4-P/m^{2}d ($n = 12$). Tilloo (2016) stated that a sudden increase of the PLR causes a decrease in the performance of the system. Similar trends were obtained in this study for the control beds. When the PLR was increased from a mean of 0.030 ± 0.003 ($n = 18$) to 0.039 ± 0.003 kgPO_4-P/m^{2}d ($n = 12$), the phosphate removal efficiency fell by 2% specifically from an average of $100 \pm 0\%$ ($n = 18$) to $98 \pm 1\%$ ($n = 12$). Nevertheless, when the PLR from the conventional systems was increased from 0.043 ± 0.004 ($n = 18$) to 0.060 ± 0.005 kgPO_4/m^{2}d ($n = 12$) kg PO_4-P/m^{2}d, the performance of phosphate removal decreased by 9% from 55 ± 3 to $45.8 \pm 2\%$. The performance of the AS-HSSFCW was enhanced or intensified by an average mean of 54, 45 and 52%, respectively, when low, medium and strong synthetic industrial WW was fed to the system. There was a significant decrease in the performance of the conventional AS-HSSFCW compared to the intensified AS-HSSFCW since gravel removes phosphate poorly and AS eliminated phosphate significantly by the process of adsorption. Tilloo (2016) studied two HSSFCWs packed with AS where one system was planted with cattails and the other system was unplanted. The respective performance of the planted and unplanted bed was 99.17 and 99.25%. For this study when the intensified AS-HSSFCW bed was fed with a medium and high PLR, a mean removal efficiency of 100 and 98% was obtained and since the operating factors (HRT and water depth) were kept the same, the main process that caused this intensification was adsorption since AS was used as the substrate. It is evidenced by this study that the adsorption mechanism contributes considerably in HSSFCW in the treatment of a WW highly loaded with phosphate.

Effect of OSLR and PLR on bed performance

The effects of OSLR and PLR on the performance of the bed were assessed statistically by Pearson correlation statistical analysis using the IBM SPPS Software version 23.0. A strong negative correlation was obtained between the OSLR and the removal efficiency of COD for the intensified coal ash HSSFCW ($r = -0.734$, $N = 30$, $p < 0.05$) and the conventional coal ash HSSFCW ($r = -0.912$, $N = 30$, $p < .05$). Hence it can be concluded that as OSLR increases, bed performance decreases. There was a very strong negative correlation between phosphate loading rate and the removal efficiency of phosphate for the AS HSSFCW ($r = -0.716$, $N = 30$, $p < .001$) and conventional alum sludge HSSFCW ($r = -0.814$, $N = 30$, $p < .001$). Therefore, it can be deduced that an increase in the PLR causes a drop in bed performance in terms of phosphate removal efficiency.

A probable reason for a decreased performance in the intensified beds when subjected to increased OSLR/PLR was due to lesser availability of adsorption sites (or more adsorption sites were occupied). However, this is by no means significant as the COD removal efficiency in the intensified bed (CA-HSSFCSW) decreased from 91% to only 85%, when the OSLR was increased from 0.238 to 1.030 kgCOD/m^2·d. Similarly, a paltry reduction of PO_4-P removal efficiency from 100 to 98% was observed in the AS-HSSFCW when the PLR was increased from 0.030 to 0.039 kgPO_4-P/m^{2}d.

K-C and K-C* model

The removal of contaminants in a HSSFCW normally follows first-order kinetics. Nevertheless, it is stated that the application of the hydraulic model is normally dependent on the aspect ratio of the HSSFCW (Von Sperling & de Paoli 2013). As the four HSSFCWs in this study had a high aspect ratio of 6:1 (length: width) supporting a plug flow regime, the mean observed COD and PO_4-P values should fit the first-order plug flow model. To insert the experimental results in the first-order plug flow model, the method of Von Sperling & De Paoli (2013) was used where a graph of mean observed COD and PO_4-P values were plotted against time and an exponential line was inserted to evaluate the R^2 value by utilising Excel Software. This shows an exponential degradation of COD and PO_4-P along the HSSFCW. Consequently, the progression of COD and phosphate obtained along the systems dropped exponentially from the inlet to the outlet of the bed.

Figure 6 shows the fit of the mean observed COD and PO_4-P values along the intensified and the conventional HSSFCWs into the K-C model at a depth of water of 0.40 m and retention time of 24 hrs for medium and strong WW.

Figure 6 clearly showed a strong correlation as indicated by the high R^2 value for the progression of COD and PO_4-P along both the intensified and control systems for the medium and strong WW. Hence, the experimental outcomes are well suited to modelling with a first-order plug flow model. This confirms the results of this study, as presented in Tables 8 and 9. It clearly demonstrates the COD and PO_4-P elimination along the systems for the intensified and the control beds followed a first-order plug flow model.

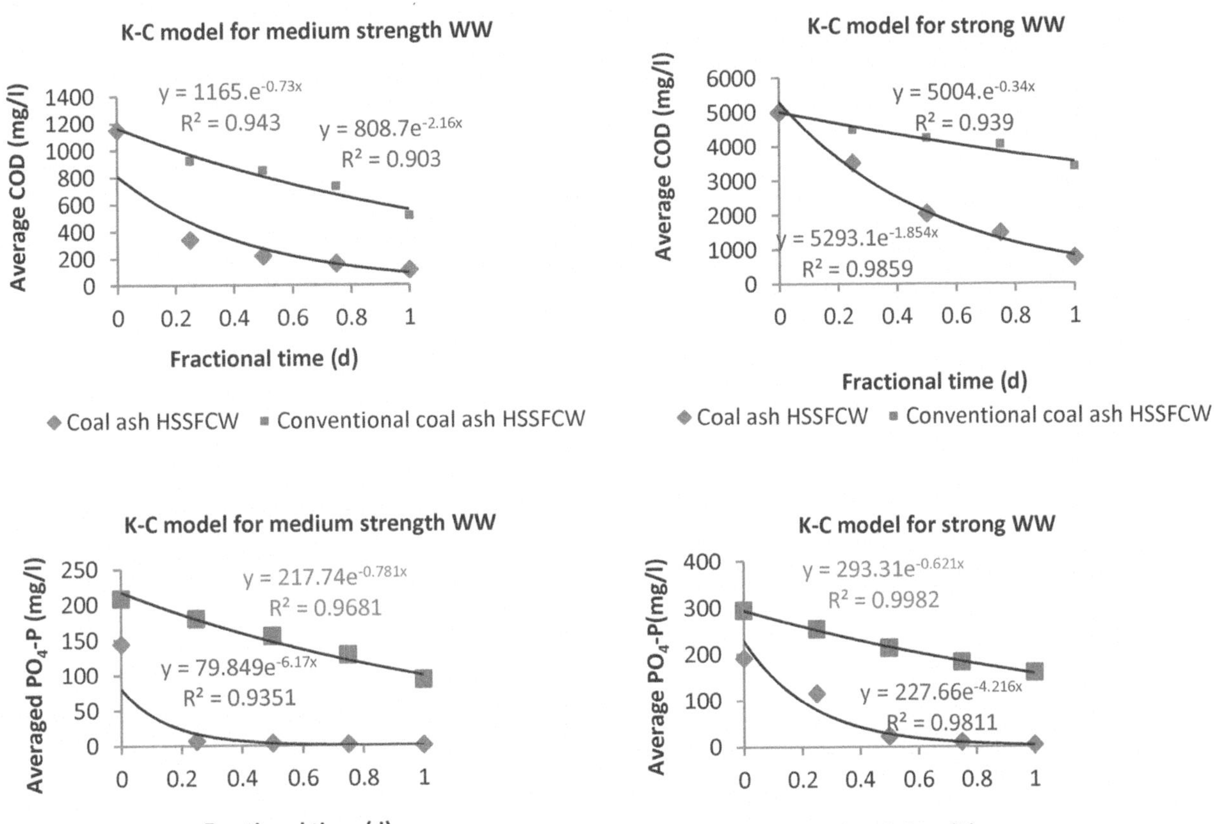

Figure 6 | K-C model for the intensified and conventional HSSFCW for medium and high-strength effluent. Note: The rate constant is higher for an intensified bed than the conventional HSSFCWs that is the intensified bed outperformed the conventional ones.

Kadlec & Wallace (2008), however, discussed that K-C model, which explained the elimination of COD and phosphate along the HSSFCW are regarded as inappropriate model since a zero COD and phosphate can only be obtained for this type of model at infinite HRT. In practice, this does not occur. The K-C model was further developed into the K-C* model, which explained a non-zero COD and PO_4-P and is represented as C*. Von Sperling & de Paoli (2013) stated that the lowest outlet COD and PO_4-P concentration will be the C*. The least COD obtained for the medium and high-strength WW were 43 mg/l and 632 mg/l for the CA-HSSFCW and 410 mg/l and 3,275 mg/l for the CCA-HSSFCW respectively. For the AS-HSSFCW, the lowest PO_4-P was 0.03 mg/l and 0.512 mg/l and for CAS-HSSFCW the C* values 78 mg/l and 138 mg/l for CAS-HSSFCW, when they were subjected to respectively medium and strong WW. The observed mean COD and PO_4-P residual concentrations were fitted into the K-C* model. Figure 7 illustrates the fit of the observed mean COD and phosphate at 24 hrs for a water depth of 0.40 m at different strength.

Trang *et al.* (2010) stated that the Kickuth design equation to evaluate the rate constant (k_{COD} and k_{PO4-P}) could be affected by some parameters such as the wetland configurations and hydraulic loadings. In view of that, other methods were proposed by Kadlec (2003), which were less influenced by these factors. This technique comprised COD/PO_4-P profiles beginning at the start to the end of the HSSFCW. Trang *et al.* (2010) detailed and implemented the K-C* model. The same method was utilised in this study to calculate the k_{COD} and k_{PO4-P}, which is explained here. The rate constants were calculated by using Equation (8).

$$k_{COD}/k_{PO4-P} = k_t.n.d \tag{15}$$

where k_t was obtained from the K-C and K-C* models, n: porosity of media used, d: depth of bed.

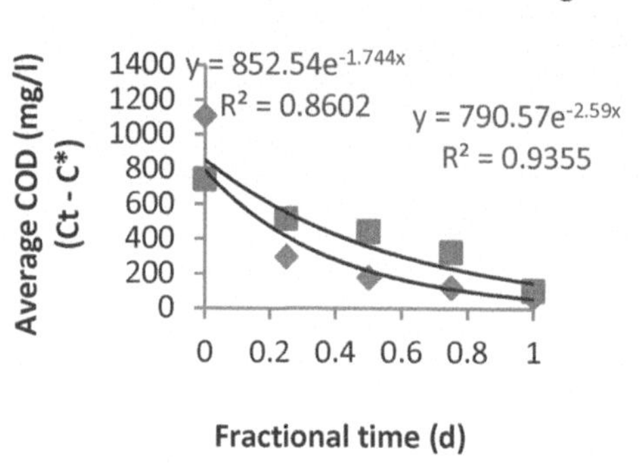

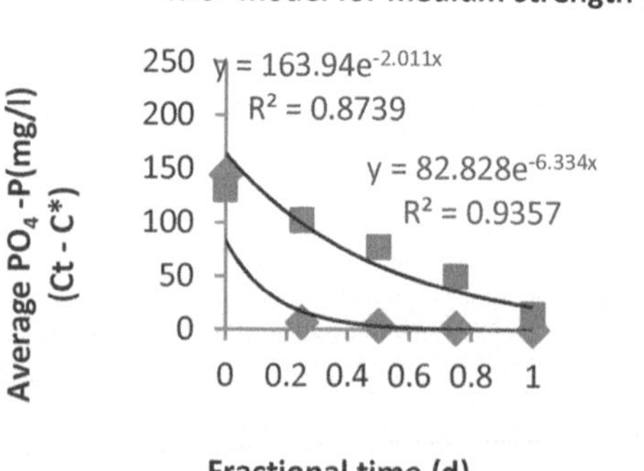

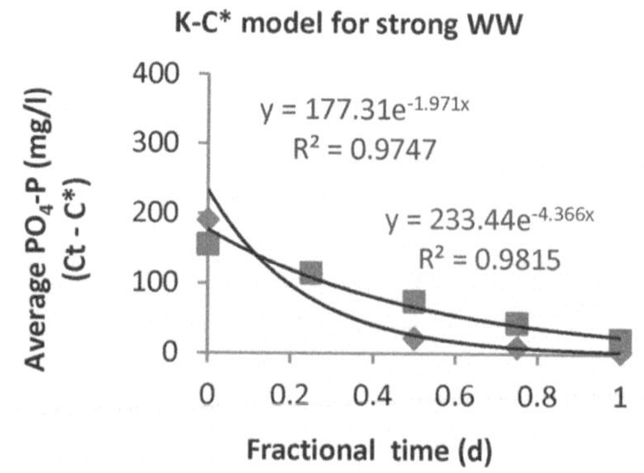

◆ Alum sludge HSSFCW ■ Conventional Alum sludge HSSFCW ◆ Alum sludge HSSFCW ■ Conventional Alum sludge HSSFCW

Figure 7 | K-C* model for the intensified and conventional beds at medium and strong strength WW. Note: The rate of constant is higher for an intensified bed than the conventional HSSFCW that is the intensified bed outperformed the conventional one.

In this study the porosity of gravel, coal ash and AS were 0.512, 0.52 and 0.46, respectively. The water depth was 0.40 m. Tables 7 and 8 illustrate the mathematical correlation between concentrations with respect to k_t, k_{COD} and their R^2 value for the K-C and K-C* models for intensified beds (CA-HSSFCW and AS-HSSFCW) and conventional systems (CCA-HSSFCW and CAS-HSSFCW) at medium and high strength.

From Tables 7 and 8, it can be deduced that the removal of COD and PO4-P for medium and high-strength WW are well described by the K-C and K-C* models as a high R^2 values were obtained for both intensified and control systems. The following finding scan be inferred from Tables 7 and 8.

• Similar fits for both K-C and K-C* model for COD and phosphate elimination at medium and high strength;

Table 7 | Relationship of rate constants of different strength effluent for the K-C and K-C* models for CA-HSSFCW and CCA-HSSFCW[a]

Model	Strength	CA-HSSFCW Correlation	k_t (/d)	k_{COD} (m/d)	R^2	CCA-HSSFCW Correlation	k_t (/d)	k_{COD} (m/d)	R^2	Increase
K-C	Medium	$C_t = 808.7e^{-2.16t}$	2.16	0.45	0.903	$C_t = 1,165e^{-0.73x}$	0.73	0.15	0.943	x3
	High	$C_t = 5,293\,e^{-1.85t}$	1.85	0.38	0.985	$C_t = 5,004e^{-0.34x}$	0.34	0.07	0.939	x5
K-C*	Medium	$C_t\text{-}C^* = 790.5e^{-2.59t}$	2.59	0.54	0.935	$C_t\text{-}C^* = 852.5e^{-1.74x}$	1.74	0.36	0.860	x2
	High	$C_t\text{-}C^* = 6,022\,e^{-3.34t}$	3.34	0.69	0.899	$C_t\text{-}C^* = 2,249e^{-2.24x}$	2.24	0.46	0.767	x2

[a]The COD rate constant indicates a faster removal of COD for the intensified bed than a conventional HSSFCW for both KC and KC* Model.

Table 8 | Relationship of rate constants of different strength effluent for the K-C and K-C* models for AS-HSSFCW and CAS-HSSFCW[a]

Model	Strength	AS-HSSFCW				CAS-HSSFCW				Increase
		Correlation	k_t (/d)	k_{PO4-P} (m/d)	R^2	Correlation	k_t (/d)	k_{PO4-P} (m/d)	R^2	
K-C	Medium	$C_t = 79.84e^{-6.17t}$	6.17	1.14	0.935	$C_t = 217.7e^{-0.78t}$	0.78	0.14	0.968	x8
	High	$C_t = 227.6e^{-4.21t}$	4.21	0.77	0.981	$C_t = 293.3e^{-0.62t}$	0.62	0.11	0.998	x7
K-C*	Medium	$C_t\text{-}C^* = 82.82e^{-6.33t}$	6.33	1.16	0.935	$C_t\text{-}C^* = 163.9e^{-2.01t}$	2.01	0.37	0.873	x3
	High	$C_t\text{-}C^* = 233.4e^{-4.36x}$	4.36	0.80	0.981	$C_t\text{-}C^* = 177.3e^{-1.97t}$	1.97	0.36	0.974	x2

[a]The PO_4-P rate constant indicates a faster removal of phosphate for the intensified bed than a conventional HSSFCW for both KC and KC* Model.

- The K-C* model has a high first-order decay rate coefficient for the intensified and control systems at medium and high strength when compared to the K-C model; and
- The rate constants for the control HSSFCW were lower than the intensified beds.

This validated the hypothesis that when implementing the process of adsorption in a HSSFCW, the performance is intensified. This also evidenced that coal ash and alum sludge are good adsorbents when compared to gravel. The intensified HSSFCW removed COD and phosphate at a faster rate than the conventional beds packed with gravels.

Furthermore, the following observations are made in this study:

- The general way in which the COD and phosphate change as the effluent moves through the wetland match with those of the literature review; and
- The removal of the observed COD and phosphate along the systems follow the K-C and K-C* models as stated by Von Sperling & de Paoli (2013), therefore confirm the data obtained experimentally.

Kadlec & Wallace (2008) also noted that the rate constant is normally hydraulic process dependent, and the same observations were made by Trang *et al.* (2010), who mentioned that the rate constant obtained from the K-C* profiles should not be used for designing as a parameter. This is because of climatic conditions, the growth of cattails and other uncertainty in the quality of water and flowrate could produce greater values than expected. Kadlec (2003) mentioned that the inlet and outlet performance outcomes from the horizontal sub-surface flow CW were used to calculate the rate constants. Consequently, the k_{COD} and k_{PO4-P} were estimated by using the results obtained from the inlet and outlet, flowrate and surface area of the system.

Equation (8) was used to compute the rate constants k_{COD} and k_{PO4-P} for both intensified and controls HSSFCW, which are shown in Table 9 for medium and high-strength WW.

It can be deduced that the rate constant computed by the models fit in the series obtained by the Kickuth equation. Therefore, the outcomes obtained experimentally are confirmed. When the strength of the WW was increased, a decrease in the rate constant was observed for both CA-HSSFCW and AS-HSSFCW as well as their controls. k_{COD} values ranging from 0.36 to 0.65 m/d and averaging to 0.46 ± 0.08 m/d and k_{PO4-P} ranging from 0.74 to 1.76 m/d and averaging to 1.23 ± 0.37 m/d were obtained irrespective of the conditions (HRT of 24 hrs and a water depth of 0.40 m) that were applied. The k_{COD} for the conventional HSSFCW was four times less than that of the intensified CA-HSSFCW one and the k_{PO4-P} was nine times less than that of the AS-HSSFCW.

Table 9 | Rate constant using the Kickuth equation

	k_{COD} (m/d)				k_{PO4-P} (m/d)				Mean Increase	
	CA-HSSCW		CCA-HSSCW		AS-HSSCW		CAS-HSSCW			
Strength	Range	Mean $\pm$ SD	Range	Mean $\pm$ SD	Range	Mean $\pm$ SD	Range	Mean $\pm$ SD	k_{COD}	k_{PO4-P}
Medium ($n = 18$)	0.39–0.65	0.50 ± 0.08	0.10–0.20	0.16 ± 0.03	0.95–1.76	1.47 ± 0.26	0.13–0.17	0.15 ± 0.01	3	10
Strong ($n = 12$)	0.36–0.43	0.39 ± 0.02	0.07–0.08	0.078 ± 0.004	0.74–1.22	0.87 ± 0.14	0.10–0.12	0.11 ± 0.005	5	8

Potential application of the intensified constructed wetland technology

The intensified constructed wetland technology addressed in this study is innovative in being developed to be applied mainly in industries in Mauritius (or other similar tropical locations such as Reunion island or Madagascar – within the region). The implementation of such an intensified HSSFCW treating high-strength WW is regarded as a sustainable treatment system where no chemicals or energy are used (Nivala *et al.* 2019). Only natural resources such as plants, gravel and by-products (AS and coal ash) are utilised. Instead of disposing the alum sludge and coal ash to landfill, by using it in CWs for treating effluent (Xu & Shi 2018; Singh *et al.* 2019) it helps to reduce the impact to landfill sites and thereby potentially increases their lifespan – this is particularly important for island countries with limited space) Alum sludge and coal ash, according to research presented here were shown to be good adsorbents for the removal of phosphate and organic matter, respectively (Babatunde *et al.* 2010; Wahyuni *et al.* 2018).

The kind of industries where this new technology can be applied are edible oil refineries, breweries, distilleries, food canning industries, tuna industries, textile industries and sugar industries among others in regions in which similar tropical climatic conditions prevail. Indeed, the objective was to simulate treatment of a wide range of medium to strong industrial WWs, taking cognisance of these industries.

Such a technology is less costly to construct and operate and is definitely more sustainable than other conventional chemical and biological treatment systems, which consume chemicals and fossil fuel-based energy and produce only sludge (Nivala *et al.* 2019). Industrial WWs in Mauritius are currently being treated by conventional physical, chemical and biological treatment systems comprising screening, settling, pH corrections, diffused air floatation, activated sludge process reactors, anaerobic sludge blanket reactors and up flow anaerobic contactors among others. In certain Mauritian industries, which cannot afford on-site treatment (due to high capex and opex) and have limited land space, their WWs are carted away, as per current local norms. Some industries in Mauritius often cannot fully afford treating these effluents through conventional physical, chemical, and biological technologies, which are usually costly. Hence such intensified HSSFCWs can be potentially applied to such industries as green technologies, which will purify the industrial effluent according to the local relevant norms.

CONCLUSION

The main research aspect of this study was to determine a range of rate constants (k_{COD} and k_{PO4-P}) for an adsorption-integrated HSSFCW treating high-strength synthetic industrial WW. These conditions were absent in literature for industries of Mauritius or other countries having similar weather conditions. The following can be concluded from this study:

- The synthetic industrial WW at different concentrations was treated by the four HSSFCWs as evidenced statistically by the paired sample T-test as the mean inlet COD and PO_4-P were different when compared to the outlet COD and PO_4-P.
- The respective COD removal efficiency for the intensified and conventional HSSFCW ranged from 82 to 96% and 29 to 62% having a mean of 89 ± 4 and $46 \pm 11\%$ irrespective of the conditions applied. The performance of the intensified bed was enhanced byan efficiency of 43% of removal of COD.
- When the COD concentration of the influent increased from 912 to 1,563 mg/l to 4,801 to 5,021 mg/l, at 24 hours HRT and a depth of water of 0.40 m, the efficiency for CA-HSSFCW decreased from 91 ± 3 to $85 \pm 2\%$ and for CCA-HSSFCW from 55 ± 6 to $32 \pm 1\%$ respectively.
- Few yellow plants were obtained in the conventional HSSFCW when the strength of the WW was increased. Nevertheless, there was no effect on the plants of the intensified beds when it was subjected to an increase in the concentration of the WW.
- The respective removal efficiency of phosphate ranged from 97 to 100% and 41 to 60% having a mean of 99 ± 1 and $50 \pm 6\%$ for the AS-HSSFCW and CAS-HSSFCW irrespective of the condition applied. An additional mean elimination efficiency of phosphate of 49% was achieved for the AS-HSSFCW.
- As the phosphate concentration increased from 316 to 380 mg/l to 488 to 520 mg/l, there was a slight decrease in the phosphate removal efficiency for the AS-HSSFCW from 100 ± 0 to $98 \pm 1\%$. While for the conventional AS-HSSFCW the removal efficiency fell from 55 ± 3 to $46 \pm 2\%$.
- Pearson correlation analysis demonstrated that the operational parameters such as OSLR and PLR had an effect on the performance of both intensified and control beds in terms of COD and phosphate. When the strength of the influent

was increased, a strong negative correlation was achieved. This showed that when the concentration was increased, the removal efficiency decreased.

- COD and phosphate removal along the four HSSFCWs fit the K-C and K-C* models, which are accepted worldwide. Hence, the COD and phosphate removal was deduced to follow a first-order reaction.
- The experimental outcomes of this study were statistically significant since high R-squared values were obtained for the removal of COD and phosphate.
- The respective COD removal rate constant for the intensified coal ash bed ranged from 0.39 to 0.65 m/d and 0.36 to 0.43 m/d for a medium and high WW and for the control beds the COD rate constant ranged from 0.10 to 0.20 m/d and 0.07 to 0.08 m/d. The PO_4-P rate constant for AS HSSFCW ranged from 0.95 to 1.76 m/d and 0.74 to 1.22 m/d for a medium and high-strength WW and for the control AS bed, the rate constant for phosphate ranged from 0.13 to 0.17 m/d and 0.10 to 0.12 m/d, respectively.

Hence, the main element of research of this study has successfully been achieved and a new range of k_{COD} and k_{PO4-P} were developed for an adsorption integrated HSSFCW using coal ash and AS as substrate.

ACKNOWLEDGEMENTS

These authors wished to thank Higher Education Commission (HEC) for offering the main author a MPhil/PhD scholarship and the University of Mauritius for provision of resources to conduct the study.

DECLARATION OF INTEREST

The authors of 'Performance intensification of constructed wetland technology: A sustainable solution for treatment of high strength industrial wastewater", Nazeemah Nurmahomed, Arvinda Kumar Ragen and Craig Michael Sheridan certify that they have NO affiliations with or involvement in any organisation or entity with any financial interest (such as honoraria; educational grants; participation in speakers' bureaus; membership, employment, consultancies, stock ownership, or other equity interest; and expert testimony or patent-licensing arrangements), or non-financial interest (such as personal or professional relationships, affiliations, knowledge or beliefs) in the subject matter or materials discussed in this manuscript.

DATA AVAILABILITY STATEMENT

All relevant data are included in the paper or its Supplementary Information.

REFERENCES

American Public Health Association (APHA) 2005 *Standard Method for Examination of Water and Wastewater*, 21st edn. APHA, AWWA, WPCF, Washington.

Babatunde, A. O., Zhao, Y. Q. & Zhao, X. H. 2010 Alum sludge-based constructed wetland system for enhanced removal of P and OM from wastewater: concept, design and performance analysis. *Journal of Bioresource Technology* **101**, 6576–6579.

Gebreegziabher, T. B., Wang, S. & Nam, H. 2019 Adsorption of H_2S, NH_3 and TMA from indoor air using porous corncob activated carbon: isotherm and kinetics study. *Journal of Environmental and Chemical Engineering* **7** (4), 103234.

Jang, J. & Lee, D. S. 2019 Effective phosphorus removal using chitosan/Ca-organically modified montmorillonite beads in batch and fixed-bed column studies. *Journal of Hazardous Materials* **375**, 9–18.

Kadlec, R. H. 2003 Effects of pollutant speciation in treatment wetlands design. *Journal of Ecological Engineering* **20** (1), 1–16.

Kadlec, R. H. & Wallace, S. D. 2008 *Treatment Wetlands*, 2nd edn. CRC Press, Boca Raton, FL, p. 1016.

Khan, S., Ahmad, I., Shah, M. T., Rehman, S. & Khaliq, A. 2009 Use of constructed wetland for the removal of heavy metals from industrial wastewater. *Journal of Environmental Management* **90** (11), 3451–3457.

Meng, X., Yuan, X., Ren, J., Wang, X., Zhu, W. & Cui, Z. 2017 Methane production and characteristics of the microbial community in a two-stage fixed-bed anaerobic reactor using molasses. *Bioresource Technology* **241**, 1050–1059.

Mirzaei, S. & Javanbakht, V. 2019 Dye removal from aqueous solution by a novel dual cross-linked biocomposite obtained from mucilage of Plantago Psyllium and eggshell membrane. *International Journal of Biological Macromolecules* **134** (0), 1187–1204.

Mora-Orozco, C. D. L., González-Acuña, I. J., Saucedo-Terán, R. A., Flores-López, H. E., Rubio-Arias, H. O. & Ochoa-Rivero, J. M. 2018 Removing organic matter and nutrients from pig farm wastewater with a constructed wetland system. *International Journal of Environmental Research and Public Health* **15**, 1031.

Nivala, J., Kahl, S., Boog, J., van Afferden, M., Reemtsma, T. & Müller, R. A. 2019 Dynamics of emerging organic contaminant removal in conventional and intensified subsurface flow treatment wetlands. *Journal of Science of the Total Environment* **649**, 1144–1156.

Rajkumar, K., Muthukumar, M. & Sivakumar, R. 2010 Novel approach for the treatment and recycle of wastewater from edible oil refinery industry – an economic perspective. *Resources, Conservation and Recycling* **54**, 752–758.

Saeed, T. & Khan, T. 2019 Constructed wetlands for industrial wastewater treatment: alternative media, input biodegradation ratio and unstable loading. *Journal of Environmental Chemical Engineering* **7** (2), 103042.

Shah, A. K., Zeenat, M. A., Abdul, J. L. & Syed, F. A. S. 2013 Utilization of fly ash as low-cost adsorbent for the treatment of industrial dyes effluents- a comparative study. *Journal of Engineering and Technology* **2** (1), 1–10.

Shepherd, H. L., Grismer, M. E. & Tchobanoglous, G. 2001 Treatment of high-strength winery wastewater using a subsurface-flow constructed wetland. *Journal of Water Environment Resource* **73** (4), 394–403.

Sheridan, C. M., Glasser, D. & Hildebrandt, D. 2014 Estimating rate constants of contaminant removal in constructed wetlands treating winery effluent: a comparison of three different methods. *Process Safety and Environmental Protection* **92** (6), 903–916.

Singh, N., Mithulraj, M. & Arya, S. 2019 Utilization of coal bottom ash in recycled concrete aggregates based self-compacting concrete blended with metakaolin. *Resources, Conservation & Recycling* **144**, 240–251.

Stefanakis, A. I., Seeger, E., Dorer, C., Sinke, A. & Thullner, M. 2016 Performance of pilot-scale horizontal subsurface flow constructed wetlands treating groundwater contaminated with phenols and petroleum derivatives. *Ecological Engineering* **95**, 514–526.

Swain, A. K., Sahoo, A., Jena, H. M. & Patra, H. 2018 Industrial wastewater treatment byaerobic inverse fluidized bed biofilm reactors (AIFBBRs): a review. *Journal of Water Process Engineering* **23**, 61–74.

Tilloo, A. 2016 *Phosphate Removal in a Horizontal Subsurface Flow Constructed Wetland Packed with Alum Sludge Using Cattails. BEng (Hons.) Dissertation*, Faculty of Engineering, University of Mauritius, Mauritius.

Tolić, K., Pavlovića, D. M., Židanića, D. & Runje, M. 2019 Nitrofurantoin in sediments and soils: sorption, isotherms and kinetics. *Journal of Science of the Total Environment* **681**, 9–17.

Trang, N. T. D., Konnerup, D., Schierup, H. H., Chiem, N. H., Tuan, L. A. & Brix, H. 2010 Kinetics of pollutant removal from domestic wastewater in a tropical horizontal subsurface flow constructed wetland system: effects of hydraulic loading rate. *Journal of Ecological Engineering* **36**, 527–535.

Udom, I. J., Mbajiorgu, C. C. & Oboho, E. O. 2018 Development and evaluation of a constructed pilot-scale horizontal subsurface flow wetland treating piggery wastewater. *Ain Shams Engineering Journal* **9** (4), 3179–3185.

Von Sperling, V. & de Paoli, A. C. 2013 First order COD decay coefficient associated with different hydraulic models applied to plant and unplanted horizontal sub surface constructed wetlands. *Journal of Ecological Engineering* **57**, 205–209.

Wahyuni, N. L. E., Soeswanto, B., Akmal, H. & Puspita, N. 2018 Effect of particle size distribution and acid treated coal bottom ash on TSS and COD removal from textile effluent using fixed bed column. In *IOP Conference Series: Earth and Environmental Science*, Vol. 160.

Wu, S., Kuschk, P., Brix, H., Vymazal, J. & Dong, R. 2014 Development of constructed wetlands in performance intensifications for wastewater treatment: a nitrogen and organic matter targeted review. *Journal of Water Research* **57**, 40–55.

Wu, S., Wallace, S., Brix, H., Kuschk, P., Kirui, W. K., Masi, F. & Dong, R. 2015 Treatment of industrial effluents in constructed wetlands: challenges, operational strategies and overall performance. *Environmental Pollution* **201**, 107–120.

Xu, G. & Shi, X. 2018 Characteristics and applications of fly ash as a sustainable construction material: a state-of-the-art review. *Resources, Conservation & Recycling* **136**, 95–109.

Yan, Y., Sun, X., Ma, F., Li, J., Shen, J., Han, W., Liu, X. & Wang, L. 2014 Removal of phosphate from etching wastewater by calcined alkaline residue: batch and column studies. *Journal of the Taiwan Institute of Chemical Engineers* **45** (4), 1709–1716.

Yang, Z., Yang, L., Wei, C., Wu, W., Zhao, X. & Lu, T. 2018 Enhanced nitrogen removal using solid carbon source in constructed wetland with limited aeration. *Journal of Bioresource Technology* **248**, 98–103.

Zhou, X., Wang, R., Liu, H., Wu, S. & Wu, H. 2019 Nitrogen removal responses to biochar addition in intermittent-aerated subsurface flow constructed wetland microcosms: enhancing role and mechanism. *Journal of Ecological Engineering* **128**, 57–65.

Zhuang, L. L., Yang, T., Zhang, J. & Li, X. 2019 The configuration, purification effect and mechanism of intensified constructed wetland for wastewater treatment from the aspect of nitrogen removal: a review. *Journal of Bioresource Technology* **293**, 122086.

Zingelwa, N. S. & Wooldridge, J. 2009 Tolerance of macrophytes and grasses to sodium and chemical oxygen demand in winery wastewater. *South African Journal of Enology and Viticulture* **30** (2), 117–123.

First received 6 October 2021; accepted in revised form 23 February 2022. Available online 8 March 2022